D0146233

THE CULTURAL MEANING OF
THE SCIENTIFIC REVOLUTION

The Cultural Meaning of the Scientific Revolution

Margaret C. Jacob

Eugene Lang College

NEW SCHOOL FOR SOCIAL RESEARCH

TEMPLE UNIVERSITY PRESS Philadelphia

For Pat Kushner and Joan Moreno-La Calle

Q
175.5
J3
1987b

Temple University Press, Philadelphia 19122

First Edition
987654321
Copyright © 1988 by Alfred A. Knopf, Inc.

Library of Congress Catalog Card No. 87-050900

ISBN 87722-536-2
Manufactured in the United States of America

PREFACE

New knowledge and fresh understanding, the goals of all research, demand constant change in the way that research is conducted. At the same time, research in any field is the expression of a cultural tradition that largely shapes the questions asked, the methods used, and the results presented. This tension between new perceptions and established standards is the heartbeat of academic disciplines, and modern universities are organized to keep that tension high. Although no longer confident that this process guarantees the progress of civilization, we sustain it because in a more limited sense it works. Year by year we command more information and learn to organize and analyze it in ever more elaborate and sophisticated ways. Specialization is the serious price paid for such gains. New information can be acquired and its validity carefully tested within a narrow focus. It is more difficult to determine how bits of new knowledge should be integrated with what was known before.

The benefits of specialization and the problems it brings have been particularly significant for the discipline of history. Over the last generation, the study of history has expanded and changed as much as any discipline in the social sciences or humanities. It has done so in large part by borrowing heavily from them, while maintaining most of its traditional concerns. The interests, theories, and methods of sociology, demography, economics, anthropology, semiotics, and literary analysis, which are readily apparent in current historical writing, have also directly affected historical research.

Thus a score of specialties have been added to the familiar ones of period and place by which historical expertise is identified. In addition, the present continues, as it always has, to stimulate new perspectives on the past. Contemporary concerns —about social order, the family, welfare, racism, women's roles, and inequality; the experience of mass culture, political oppression, and demands for democracy; the fact that European power does not dominate the world as it once did; and increased knowledge of non-Western societies—have led historians to pose fresh questions about European history. The historical study of

Europe is also now more international in the double sense that historians of many nations study the same problems and that they are less inclined to confine those topics within national boundaries.

So the discoveries and successes of one historical field quickly influence work in others. The emphasis in this generation upon social history has enabled us to learn more than ever before about all eras of European history and in particular about the attitudes and lives of ordinary people with little power. By definition, this type of social history explores detailed knowledge. Its larger significance requires that the disparate facts about daily life, local behavior, or the activities of unorganized groups be related to some general view of social relations and patterns of change. Theory is essential to analysis.

As historians mine old archives for new information, the meaning of which may only be revealed through complicated techniques of cultural or quantitative analysis, they do so because they have posed fresh questions. Minute research and (and at least implicit) theory must be combined to create a history that integrates new knowledge of ordinary life with the larger fabric of historical interpretation. Discovering the significant patterns and establishing a coherent perspective that can encompass both new findings and what was known before remain the great challenges of historical study. The search for such a synthesis is the very essence of historical writing. That is not new, but in the past this quest was expressed primarily through a powerful rhetoric redolent of larger purpose and was organized in sweeping narratives that gave to chronological sequence the aura of cause and effect. Newer methods and more specialized knowledge have made those established devices more difficult and less satisfactory.

The modern synthesis is likely to move on several levels at once. It must define an important problem or set of problems, using evidence both to define them and to establish their significance. These problems must then be carefully placed in historical context. The modern synthesis, which must relate the latest findings to the relevant theories, is also expected to address the scholarly debates from which central questions derive. Thus coherence is less likely to result from narrative flow or rhetorical values than from a consistent interpretative argument. When well written, as of course it should be, the carefully elaborated historical synthesis can nevertheless join with the best of older historical writing in conveying the color and fascination of a way of life now passed and the intellectual challenge of seeking to understand a particular society.

The books in this series present that sort of modern synthesis. The excitement of recent research, with its diverse topics and methods, makes this an excellent moment for such a publishing venture. And historical synthesis, which must operate on several levels at once, offers the advantage that it can speak to many kinds of serious readers—to freshmen discovering the patterns that make sense of the panorama of Western history, to advanced students interested in the meaning of historical context and important scholarly debates, to scholars ready to wrestle with a new perspective offered by a distinguished colleague, and to anyone who enjoys thinking about how to understand significant historical developments. The authors of these books have accepted the challenge of attempting in a single volume to accomplish so much. Active scholars building upon their own research, these authors set forth arguments that specialists will need to consider. Their assessment of recent findings and of the general scholarly literature will be especially helpful to graduate students and advanced undergraduates. Their perspectives on important aspects of European history should prove useful and stimulating to beginning students in introductory courses and to their teachers who wish to integrate current research into an established framework. In short, different readers can all appreciate the same well-written and intelligent book whether initially drawn to its broad picture of an important epoch, the scholarly debate of which it is part, or the new findings it presents—and that remains one of the glories of historical study even in an age of specialization.

Science is obviously central to modern Western civilization, but historians have had trouble in explaining how it got there. Many different histories are relevant. There is the history of great ideas about natural phenomena, the history of what scientists have done (which quickly divides into a series of separate histories of each branch of science), the institutional history of how science came to be supported and scientists produced, the history of how science has been understood and used by government and business. It took a particular kind of society to make of science a distinct and dynamic enterprise. Special circumstances shaped the multiple and complicated intersections between different sciences and various parts of society. The roles of the church, the state, and the economy were obviously critical. Exploration of these issues has highlighted the important question of the relationship between science and industrialization, a subject on which there has been much debate. Given the growth in the history of science as a separate field of study with innumerable scholarly accomplishments, given increased understanding of economic history, and given the rich new work on the social and cultural history of Europe, can we now achieve a fuller understanding of the place of science in European life?

In this fascinating and significant study, Margaret Jacob presents a fresh perspective which shows that we can. She focuses directly on the disputed issue of the links between science and technology, but she treats it as a problem of cultural process. She shows how scientists made science responsive to the needs and interests of local groups that included artisans and entrepreneurs, and she shows that science became important to elites when it was seen to be compatible with their cultural values and political interests. Philosophy, attitudes, institutions, and economic pressures met in social practices that established the importance of science. Europe's scientific revolution is thus seen in broad historical perspective as the result of intellectual, cultural, and social developments rather than as the predictable achievement of a remarkable cluster of geniuses. Jacob's interpretation of one of the major changes of modern history, presented in a sophisticated synthesis based on close analysis and original research, using current work in many fields presents a picture and forms an argument that in its clarity, telling detail, and comprehensiveness will hold the interest of students and experts. Whether they approach it as a striking summary of the latest work, a fascinating account of science and industry, or a stimulating challenge to older views, readers of *The Cultural Meaning of the Scientific Revolution* will share in the excitement of historical research and the new perspectives it can bring.

RAYMOND GREW
Ann Arbor, Michigan

ACKNOWLEDGMENTS

Various universities and libraries contributed space and expertise to this project. The University of Amsterdam and its excellent department of science dynamics (*Wetenschapsdynamica*) gave me an intellectual home in 1983–1984 that will not easily be forgotten. The vigor and enthusiasm of my colleagues there sustained my inquiries into Dutch history. Their criticisms and suggestions were always enlightening and in various places have found their way into the text. The British Library and many local record offices in Britain and The Netherlands also assisted, and I especially wish to thank Norman Leveritt, John Hamell, A. Clark of The Royal Society, and the librarians at the University of Amsterdam and the Institute for the History of Science at Utrecht.

The City University of New York gave me financial support as well as leaves of absence. Its Research Foundation contributed materially to the research and completion of the book. Grants and fellowships must be gratefully acknowledged from the American Council of Learned Societies, the Fulbright program, the National Science Foundation, and the Clark Library, Los Angeles, in the form of a summer directorship of graduate study. The University of Leiden hosted me during my semester as a Fulbright scholar. Not least, the New School for Social Research gave me the necessary time to complete the writing of this book in the summer of 1985. Robert Rivelli typed large portions of the final draft. His recent and untimely death leaves his many friends at the New School very saddened.

One teacher and friend, Henry Guerlac (d. 1985), could not witness the completion of this project. Some of the themes here explored first presented themselves in his Cornell seminars. That these questions have stayed with me for so many years and continued to intrigue is testimony to what an extraordinary teacher he was.

Various people read various portions of this text; none of them should be blamed for any of it. I am very grateful to Joyce Appleby, Vincenzo Ferrone, Wijnard Mijnhardt, P. W. Klein, Thomas Hughes, J. G. A. Pocock, Jonathan Barry, Roy Porter, and the editor of this series, Raymond Grew, who was especially diligent. J. R. Ravetz kindly sent his comments in the very final stages of revision. Seminar participants at the Fondazione Einaudi, Turin, the Clark Library, and Eugene Lang College always sent me away stimulated and mindful of revisions. Some of the ideas in this text first took shape in a joint paper written in 1981 with James R. Jacob. His ideas and writings have contributed then and now, and they are gratefully acknowledged, particularly in chapters one and three.

Friends somehow make these projects memorable and their successful completion joyful. They provide everything from wisdom and solace to a roof over one's head in far away places. For all those moments of friendship I wish to thank M. Boesveld, S. Kennedy, G. Pheterson, G. Mandersloot, K. Ashfield, J. Appleby, W. Klauss, D. & H. Koenigsberger, J. & G. Roche, P. Mack (especially), C. & B. Hill, F. Carroll, R. Perry, and S. Leydesdorff. This book is dedicated to two friends, Pat Kushner and Joan Moreno-La Calle. Since our college years, Pat has given encouragement, met planes, stored research notes and always approved. Joan put up with the writing stage of this project with remarkable good cheer.

MARGARET C. JACOB
The Institute for Research in History

New York, June 1987

CONTENTS

THE CULTURAL MEANING OF
THE SCIENTIFIC REVOLUTION

INTRODUCTION

This book seeks to explain the historical process by which in the seventeenth and eighteenth centuries scientific knowledge became an integral part of Western culture. That simple statement may strike the student, at least at first glance, as inherently paradoxical. Is not science such an essential part of our understanding of the world that its preeminence within our culture may be assumed as simply basic to the Western mind from the ancient Greeks onward? Yet by the late seventeenth century the science of the ancient Greeks, especially as embodied in the writings of Aristotle, and later Ptolemy, was largely discredited within elite culture. The ancient understanding of the natural world bears little or no relation to our own, and our own is not significantly dissimilar to that of the eighteenth century.

The science that became an integral part of our culture by the middle of the eighteenth century rested on certain philosophical and mathematical innovations that in very general terms can be dated fairly precisely, from the publication in 1543 by Copernicus of his *De revolutionibus orbium coelestium (On the Revolution of the Heavenly Orbs)* until the publication in 1687 of Newton's *Principia (Philosophiae Naturalis Principia Mathematica)*. This science was a very different science from that found in other cultures (now or then), and it was based largely on the actual observation of visible bodies in motion in the heavens and on earth. It required that the results of observation be described largely according to mechanical principles (that is, contact action between bodies) as well as mathematically. This science was in turn deemed more acceptable than its predecessors, in part because it was capable of direct application to specific industrial and technological needs, to the achievement of an unprecedented impact on, and control over, the natural environment. The methodology of the new science came to rest on the interaction of experiments and hypotheses producing probable explanations, in the end laws of nature that

in turn could be applied to other phenomena, previously unex-
plored. The progressive quality of this new science lay in the
permission it gave the inquirer to alter hypotheses, whether as a
result of experiment or intuition, and in consequence through
trial and error to alter the laws themselves or to invent new ones.
This science has usually been described in the history books as
the language, methodology, and discoveries—in short the
achievement—of the Scientific Revolution.

Because we are seeking here to understand the integration
into Western culture of that achievement, a process begun in the
early seventeenth century and largely completed by the early
nineteenth century, certain traditional problems need not, in the
first instance, concern us. In their place are put other questions
traditionally avoided by most histories of the new science. In
general we are not trying to explain why or how Copernicus,
Galileo, Newton, or a host of now less famous natural philosophers (to use the term they would have understood) conducted
the experiments they chose to pursue, or solved the mathematical problems that consumed their interest; how, in short, they
came upon their discoveries. Many excellent historical accounts
of their scientific development now exist (see bibliographical
essay), although a number of those accounts incorrectly presume
that a history of mathematics or experimental methodology is
sufficient to explain the cultural meaning given to the new science, the phenomenon with which we are here concerned.

For our purposes we may presume upon the genius displayed
by some of these inventors of the new science, or upon their
sheer doggedness in pursuit of a new understanding of nature.
But genius or perseverance in itself cannot account for the place
achieved by or the meaning assigned to science in our culture.
Had the integration of science depended largely on genius, we
might have found its possessors as court mandarins presiding
over a special wisdom enjoyed by a select few, or as cloistered
academicians preaching to the few thousands in any country in
any decade who made their way to an early modern European
university; or, in a darker scenario, as itinerant heretics in search
of patrons, or as suspect prophets subjected to house arrest,
imprisonment, or death by burning. None of those fates awaited
the western and northern European promoters of the new science. Only Galileo after his confrontation with the church in

1633 languished under house arrest in Florence (see Chapter 1), although some of his European followers in the new mechanical philosophy did, at moments, fear a similar fate (see Chapter 2).

That they escaped from ever experiencing their darkest fears cannot be explained by reference to their genius, however real it may have been. Larger historical, that is contextual, factors account for the integration and use of science in our culture. Such social and political factors affected the prophets and progenitors of science just as they impinged on all forms of mental activity in the early modern period. Of course, it might be tempting to assert that those factors also, and simply, created that very science, the cultural integration of which we seek to explain. My intention is not to encourage that assertion.

The approach taken here, which somewhat separates scientific discovery from its assimilation, possesses another danger. I do not wish to imply that the new science was some foreign object in need of assimilation, like snow in the desert, and that its genesis is ultimately mysterious, the work of incomprehensible genius, and worse still, its origin simply irrelevant to history. The new science, that systematic study of the natural world which presumed matter in motion to be the central object of inquiry, was itself a cultural artifact. All expressions of human creativity are historical—that is, bound by time and place and, as such, relative to eras and epochs. There is no absolute ahistorical Truth over which a select clergy, however much it might seek to do so, may exclusively preside.

The approach to science taken here permits the presence of rationality in science (as in music or in other expressions of human ingenuity). To deny that rationality seems to bring us dangerously close to cultural nihilism, as well as to the callous disregard for the scarce material resources non-Western peoples have been willing to commit to obtain our science. Yet the origins of rational mathematics, or the genesis of a new method for replicable experiments whether in astronomy or optics—however important and fascinating—do not concern the general historian in search of a brief, but coherent account—one comprehensible to general readers—of how, by the eighteenth century, science came to be such a vital part of our culture.

Certain assumptions are basic to this study. It assumes that science possesses a social and cultural—that is, a nonmathematical or technical—history that needs telling, however briefly. It further presumes that the new science did not become an essential, perhaps the essential, part of our culture largely because educated, literate Europeans in the seventeenth and eighteenth centuries perceived its self-evident truth and in that moment of illumination embraced it. We cannot write a history which presumes that the educated Western elite and the scientists they read possessed an inherent rationality that was superior to all previous or contemporary rationality, which in turn made their new version of the natural world obviously correct. This is not to deny the presence of rationality and progressive discoveries in the early development of modern science. Rather it is to assert that in order to perceive a body of knowledge as rational—and eventually, by the eighteenth century, as superior to its competitors—a complex set of historical circumstances had to exist that enabled some people, and eventually a whole culture, to embrace a new understanding of the natural order and to apply that understanding in ways that increasingly came to be seen as beneficial to themselves. Those circumstances can be discovered in the written records of the past, in the language used by natural philosophers and their listeners alike. In the shared language of the natural philosophers and their audience we find the key to unlocking the historical process by which scientific language came to be one of the central vocabularies in our mental world.

In early modern Europe the new science was assimilated and used often as a result of certain larger historical developments: the Protestant Reformation of the sixteenth century; the invention of printing; the development of centralized national states and their attendant bureaucracies; the fear, or reality, of political revolution or social disorder coming from the illiterate or less-educated mass of the population (see Chapter 3); and not least, by the late seventeenth century in western Europe, the desire on the part of the landed elite to transform the basis of their power away from feudal dominion over large land holdings and a dependent peasantry, into a more flexible use of their resources and power. The necessity of sharing political and economic power with urban and mercantile elites forced both the landed and the mercantile to seek new means by which they might explore and exploit the world around them. The presence of scientists or scientific devotees trained in the schools and universities, but hostile to the monopolies enjoyed by the clergy and the scholastic learning they promoted, made for the possibility of a unique merger between the interests of natural philosophers and the needs of the literate and propertied elites (see Chapter 4). Where that merger was effected successfully, first in England then in France and the Low Countries, a synthesis of scientific knowledge and material interests led to an unprecedented conquest of nature through commercial and industrial development. The road from the Scientific Revolution to the Industrial Revolution, although occasionally requiring detours or impasses, is more straightforward than we may have imagined (see Chapters 5 through 7).

Because the very language of science, namely natural philosophy, depended on such advanced forms of literacy, both for its articulation and for its assimilation, we are justified in confining our attention to certain select elites: university trained philosophers, wealthy aristocrats, and merchants and clergy who could afford books on science—in short, probably to no more than 5 percent of the male population and only an infinitesimal percentage of the female population. Of necessity such a small percentage of the population was concentrated in areas of advanced literacy, for the most part in the cities of northern and western Europe: London, Amsterdam, Paris. Consequently in any general account of the assimilation of the new science, England and The Netherlands, both north and south, as well as northern France, must figure prominently because in those places we find the largest concentrations of wealth and literacy. As we shall see, the history of the assimilation of science in those various local settings differs enormously. It was contingent, as we have come to expect, on the vicissitudes of history, on religion, politics, and economic development. In this book when we speak of elites we have very distinct images in our historical imagination. We imagine, for example, a landed gentleman in one of the southern counties of England who voted in parliamentary elections, perhaps even stood for Parliament, and expected the local clergyman to preach in the pulpit in ways amenable to that gentleman's interests. Or we imagine a prosperous urban-based merchant, in London or Amsterdam, who read books, generally of a religious nature, and could follow philosophical argumentation. Always we include educated clergymen, Protestant or Catholic, as well as lawyers, doctors, journalists, city magistrates, and men suffi-ciently educated, generally at universities, to take their place as tutors in households or as minor officials at court. Army officers —such as the young natural philosopher Descartes—were also frequently well educated, as were many members of the aristoc-racy from which they generally came. Seldom did women figure publicly in any of these groups, but that did not prevent privately educated women (women were not permitted to attend universi-ties) from following natural philosophical discussions and even, by the eighteenth century, contributing to them. Our cast of historical characters is, relatively speaking, very small; but it me-diated over a body of new knowledge as important in our culture as any ever articulated, before or since.

The approach to science and its cultural meaning taken here breaks with an older genre of writing about science in history that concentrated exclusively on "pioneers," "achievements," "prog-ress." As early as the mid-eighteenth century, that heroic account of the progress of Western science from genius to genius, from breakthrough to discovery after discovery, had become the stan-dard explanation for its success among its champions and pro-moters. It is a nice history, edifying and uplifting but not very profound. In the postwar era some philosophers of science (a relatively new academic discipline) tried to elevate that narrative account of the triumph of science to a new status. They argued that science is "rational" and that which is social or cultural is "irrational." Hence there is no need to write the kind of cultural history of science in society presented here, or so the argument implies.[1]

If we were to accept this distinction between science and history, between the "rational" and the "irrational," we would render the historical process, in this case the acceptance and use of scientific ideas—the cultural meaning of science—into an area fraught with terrible methodological difficulties. How could we write historically, laboring with such presuppositions? Happily for students and teachers alike, those philosophers who hold to such assumptions seldom write history, for understandable rea-sons.

But those who do write it, especially in brief form, leave out much of the complexity of the human past. If you approach this narrative as a series of general lectures you may be more likely to forgive the omissions and seek through further reading to augment them. Yet the reader will be confronted here with a story that arouses moral judgments and value-laden pronouncements. Perhaps no single area of human inquiry now provokes greater passion than does the cultural meaning of science. In the late twentieth century we approach its history, that is, the history of our culture, with the realization that we may have assimilated that which has become capable of destroying us. In consequence there are moralists among us who would judge science as dangerous in itself. Perhaps we might ease our predicament by turning our moral and critical faculties to the scrutiny of the historical process by which we got to where we are (see epilogue). History, if it is to be a humanistic discipline, must enlighten, especially where it can neither console nor condemn.

NOTES

1. The literature to which I refer is vast, and frequently its terminology, not to mention its polemical assertions, are incompatible with historical thinking. For an example, see Imre Lakatos and Elie Zahar, "Why Did Copernicus' Research Program Supersede Ptolemy's?" in Robert S. Westman, ed., *The Copernican Achievement* (Los Angeles: University of California Press, 1975), pp. 354–357. For a balanced approach to the questions raised by postwar philosophy of science, see Steven Lukes, "Comments on David Bloor," *Studies in History and Philosophy of Science,* vol. 13, no. 4 (1982), pp. 313–318, with reference to other very stimulating essays in the same volume by David Bloor, Mary Hesse, and others.

CHAPTER 1

Laying Claim to an Audience: The First Prophets of the New Science

Why then do we hesitate to grant it [the Earth] the motion which accords naturally with its form [that of a sphere], rather than attribute a movement to the entire universe whose limit we do not and cannot know? And why should we not admit, with regard to the daily rotation, that the appearance belongs to the heavens, but the reality [of it] is in the Earth?

Copernicus, *The Revolution of the Heavenly Orbs,* 1543

THE SOCIAL CONTEXT OF COPERNICANISM

In 1543 Copernicus (1473–1543) asked, somewhat rhetorically, in his highly technical Latin treatise on the motion of the planets, why the notion that the earth moved in the universe was resisted so adamantly by his contemporaries. Of course he knew perfectly well why they resisted, why they placed their faith in centuries of learning that located the earth in the center of a closed universe, surrounded by luminous bodies, planets visible because of their light but not materially real. The Ptolemaic or geocentric system that located the earth in the center of the universe also, and quite simply, worked reasonably well. Its mathematics were excessively complex, but it could predict the positions of the planets. To dislodge it would require what Copernicus could not have imagined. An enormous transformation in mentality would be necessary, taking two hundred years to complete, before heliocentricity, and all that it implied, gained widespread acceptance among the educated elites of Europe.

Among the educated elite in sixteenth- and seventeenth-century Europe we can identify certain key groups whose acceptance or rejection of the new science would set its fortune. Princes and their courts offered the possibility of patronage or, more important, the protection and promotion of new ideas as compatible with their power. Among these groups could be included the pope, to whom Copernicus dedicated his treatise in 1543, or the mercantile aristocracy of the Italian city-states, whom his most important follower, Galileo, wooed over a half-century later.

Yet by the early seventeenth century new men (and some women) had joined the select company of the highly literate. Wealthy merchants bought and read books. They also applied simple mathematics to day-to-day transactions; they weighed goods and kept accounts. To their "common sense" Galileo— also the most famous Copernican of his time—appealed when he argued that his geometrical approach to physics took into account the real world of everyday material objects. When errors crept into his physics, he explained to his readers, it was because the experimenter was like "a calculator who does not know how to keep proper accounts."[1] Galileo presumed that it would be largely unthinkable for his abstractions not to bear relation to reality, just as it would be bizarre for "computations and ratios . . . not to correspond to concrete gold and silver and merchandise."[2]

As Galileo was to discover, both merchants and princes were less skilled in abstraction than their teachers and preachers. The clergy, whether Catholic or Protestant, were the purveyors of the written and spoken word. They controlled all the universities and pulpits; learned discourse, and hence the very language of natural philosophy, had been their province for centuries. When the well-educated clergyman spoke from the pulpit he translated complex metaphysical assumptions about the universe and its relation to the deity into the daily language of religious piety. If the clergy could not—or would not—make that translation, then the language of natural philosophy, in short the language of science, stood separately from that of commonplace religiosity.

Such a separation between the religiosity of the people and the beliefs of the learned did indeed become commonplace by the late eighteenth century. In the decades after the death of Copernicus, however, the advocacy of that separation was greeted with suspicion, especially by the Catholic clergy. To put their suspicions in perspective and not to derive unwarranted conclusions from them, we should remember that within a hundred years—from, let us say, the 1590s to the 1690s—the new science would be championed by other Christian clergy and laity alike. The point here is that there was no inherent and inevitable warfare of science and religion in preindustrial Europe. There were, however, cultural moments, such as occurred in the time of Galileo, when the restructuring of certain basic assumptions about the universe proved profoundly threatening to religious orthodoxy.

In this chapter we are exploring the earliest stage of the historical transformation by which the new science became an inherent part of the Western understanding of nature. In that effort we must examine the ideological content of the earliest appeals made by the first prophets and promoters of the new science. By the early seventeenth century they had ceased to write exclusively in Latin. Quite purposefully the followers of Copernicus, in particular Galileo, wrote in Italian (or simply in the vernacular), and in so doing, they appealed not only to the clergy but, more dramatically, to the educated laity in the urban commercial centers as well as to the new monarchs and the salons of the princes and aristocrats. At every turn these vernacular appeals sought to integrate science into the values and interests of the elites to whom they primarily spoke.

Some of the themes found in the appeals of Galileo can also be found in the writings of Copernicus and those of his earliest followers. In the preface to *De Revolutionibus* (1543) Copernicus revealed that he knew that some of his critics would attempt to repudiate him along with his "insane pronouncement" that the sun lay at the center of the universe. Throughout his text he pleaded a case that was mathematically and rhetorically elegant. He also made clear his personal piety as well as his dedication to the natural philosophy of Aristotle. That philosophy, as vastly modified by medieval scholasticism, stood as the commonplace wisdom of university teachers and preaching clergy alike.

The legacy of medieval Aristotelianism is one of the most difficult to assess. Once discarded, any body of thought acquires the aura of the arcane, if not the moribund. By the seventeenth century new scientists excelled at the sport of mocking the schoolmen,* the promoters of Aristotelianism. Those strictures

---

*See glossary of terms.

have frequently been repeated by historians, and in consequence, a balanced picture of the Aristotelianism taught in the universities at the time of Copernicus or Galileo is difficult to achieve. What is clear from recent scholarship is the emphasis placed by the followers of Aristotle on an empirical tradition. Simple observation was highly valued, more so, in some cases, than mathematics. So, too, was logic. Argumentation took the form of hypotheses tested by syllogisms. Among the scholastic disciplines, philosophy and theology ultimately took precedence over any of the other sciences. To draw out the full implications of all those assumptions for the new science will require an examination of the most dramatic example of its conflict with Aristotelianism given by that age. The case of Galileo and heliocentricity, ostensibly versus Ptolemy and the followers of geocentricity, provided such a test of the flexibility of Aristotelianism in the face of a potentially challenging body of knowledge.

Decades before Galileo's confrontation with the church, one of Copernicus's own pupils assessed the fundamental issues raised by his mentor's astronomy. Rheticus acknowledged that to say that the sun resides in the center of the universe contradicts certain passages in Scripture. There is no way around that problem other than to divorce this new astronomy from the common language of religious discourse. This pupil of Copernicus argued as early as 1540 that the Bible should be seen as adapting itself to vulgar opinion. It was an argument for the separation of Scripture and science that Galileo would later adopt, to his peril. In addition, Rheticus saw the potential for conflict with the Aristotelian schoolmen. They wanted to deal with Copernican propositions as hypotheses—no more and no less. Rheticus puts his objection to their methods succinctly and defiantly: "The philosophers say that some things are known to nature, but unknown to us. To this category let us indeed consign also disputes about hypotheses."[3] Disputing about hypotheses lay, however, at the core of the scholastic curriculum. Within that form of disputation content unfolded. However frequently sixteenth-century university professors of physics might teach Copernicus as one more hypothesis—if only for refutation—such practice avoided the issues raised by Rheticus. It also could never encompass the full-scale assault on scholasticism and the old science championed by Galileo. We should remember that as a student, Galileo learned the astronomy of Copernicus from his Jesuit teachers in Rome, who taught it in order to refute it.[4]

The first major assault on the old science occurred in early seventeenth-century Florence, the commercial center of Tuscany. There the humanistic culture of the Renaissance had already sought to render knowledge meaningful to civic life, to the political and economic interests of aristocrats and merchants alike. By the early 1600s approximately one-third of the Florentine citizenry was engaged in some capacity in the cloth trade, or in manufacturing. In the previous century the Florentine humanists had also made their city the center of late Renaissance culture. The Florentine elite was both commercial and cultured, while its clergy ranged from popular preachers and ordinary university academicians to archbishops and cardinals of considerable theological sophistication. Within this social setting the drama between Galileo Galilei (1564–1642) and the Inquisition of the Roman Catholic church unfolded.

The confrontation between Galileo and the church became a symbol in its age, and well beyond, of the supposedly inevitable conflict between the new science and traditional Christianity. But that conflict was largely the result of historical circumstances. The Protestant Reformation of the sixteenth century had placed the leadership of the church at the center of doctrinal confrontation with Protestant theologians, as well as with "heretical" intellectuals of the period, many of whom were desperately seeking a way out of the impasse created by the irreconcilable split between Protestants and Catholics. By 1600 the church saw enemies on every side: Protestants, strong particularly in northern and western Europe, in possession of their own universities, and even dominant in certain cities and states; skeptics, hostile to doctrinal orthodoxy of any kind and found most commonly among the lay elite, particularly in France; and not least, new and heretical philosophers, frequently from a clerical background, who sought to revive the religiosity of the ancient pagans as the new foundation upon which some sort of new universal religion might be constructed. One such prophetic philosopher, the Italian Dominican Giordano Bruno, traveled to the major courts of Europe proclaiming this pagan naturalism with all its magical associations as the alternative to the doctrinal orthodoxies of both Protestants and Catholics.[5] In 1600, by order of the Inquisition, Bruno was burned at the stake in Rome; but his ideas, to which we shall return shortly, did not die with him. As a result of these challenges to its authority the church took the arbitration of theological matters out of "the community of scholars and conferred [it] on a bureaucratic institution in Rome"—namely, into the hands of the clerical administrators of the Inquisition. This shift away from the community of scholars as the final arbiters in doctrinal matters—a process well under way by the early seventeenth century—provides the context within which Galileo's condemnation in 1633 occurred.[6]

Without this context, that condemnation and its effects become extremely difficult to understand. To comprehend even more fully the events leading up to 1633 we must also explore the link that Galileo attempted to forge in Florence and the other Italian city-states between the educated laity and the new science. Operating in the tradition of civic humanism, as well as spurred by the power of the Inquisition and its clerical supports, Galileo sought allies among aristocrats and merchants. He argued that science was uniquely suited to their interests as well as to their education and intelligence.

Given what historians now know about the formation of a distinctive elite culture in the early modern period, we would have to conclude that Galileo was among the first scientists to appeal to what was in fact a newly empowered literate culture, and, more important, that he had a reasonably accurate grasp on its values and assumptions. What he did not reckon on was the immense power of the Roman Inquisition. He may also have failed to realize how out of touch and unsympathetic the local bureaucracy of the Inquisition had become to an intellectual tradition of free scientific inquiry that had once flourished in the late medieval universities and that had never intended, nor in fact constituted, a threat to the doctrinal foundations of Christianity.

Galileo's friends in the scientific academy of Florence also believed it would be possible to win the church over to heliocentricity, to place the new science at the center of its learning. Galileo may have shared their preoccupation and, as it turned out, their miscalculations.[7] Certainly Galileo attempted to argue the Copernican case not only on scientific grounds but also on theological ones. In so doing he embarked on a very risky course, that of appealing to his elite audience in Italian while at the same time arguing as a layman on matters in which theologians held strongly and professionally sanctioned opinions.

These tensions and misconceptions within the church, and between Galileo and the church, suddenly brought the new science out of the domain of the universities and the learned disputations of the natural philosophers, and to the attention of educated Europeans. What might have remained matters for debate among experts—for example, the relative merits of the Copernican system in relation to the geocentric system of Ptolemy, or the possibility of reconciling heliocentricity with the teachings of Aristotle—now became topics of widespread intellectual interest. With Galileo's writings and his subsequent trial in 1633, the process by which the new science became integrated into Western culture began in earnest.

GALILEO'S CONFRONTATION WITH THE CHURCH

Even by the late fifteenth century, before the time of Copernicus, students at the traditional universities of Europe, whether in Florence, Cracow, Oxford, or Paris,[8] would have found themselves being taught a traditional Aristotelianism, to be sure; but they would also have found a growing emphasis on a more practical, less abstract, science: on astronomy for navigation, on the principles of simple mechanics, on metallurgy, on mathematics, and even on practical anatomy as taught at Bologna, for example.[9] Where they also attended universities influenced by the civic humanism of the Renaissance they would have learned of the necessity for the educated laity to use their learning in the service of the polity. Similarly by the late sixteenth century, if not earlier, and particularly in Protestant German universities, students learned their Aristotle along with their Copernicus. In 1592 a professor at the University of Nuremberg wrote in his lectures: "Now, just as everyone approves the calculations of Copernicus . . . so everyone clearly abhors his hypotheses on account of the multiple motion of the earth. . . . If one retains the supposition of Ptolemy, one achieves the same goal that Copernicus attained with his new construction."[10] In other words, it was possible by the late sixteenth century to accept aspects of

Copernican mathematics and still be a supporter of geocentricity, or even to be a Copernican and still believe an essentially Aristotelian explanation of the nature of matter, or motion, or the fixity of the stars in a closed universe. It seems reasonable to conclude that by the 1590s portions of Copernicus's astronomy might have gained widespread acceptance in the schools of both Catholic and Protestant Europe as part of a more mathematical and empirical emphasis within scientific education. Such a limited acceptance would not have engendered a rapid and violent confrontation with Aristotelian natural philosophy and its upholders.

When in 1632 Galileo published his appeal to the merchants and urban aristocrats of Florence and beyond, that confrontation did occur—first and most dramatically in Italy, and eventually all over Europe, both Catholic and Protestant. In the course of the seventeenth century it pitted the followers and teachers of Aristotle, for the most part clergymen, against the largely lay, but occasionally clerical and often Protestant, followers of a mechanical philosophy of nature. Both philosophies are clearly at the root of the conflict between Galileo and his clerical opponents. Added to that purely philosophical disagreement came Galileo's proclamation (first noted by J. R. Jacob) that the new mechanical philosophy forms the basis of a privileged learning suitable only for the educated few. To express this appeal to elite culture in the language of Galileo, science is fit only for "the minds of the wise" and not for "the shallow minds of the common people."

This knowledge fit for the elite entailed more than an acceptance of Copernican heliocentricity. It raised problems with Scripture and the Aristotelians that were difficult but not insurmountable. Galileo also proclaimed heliocentricity as the necessary corollary of new philosophical assumptions antithetical to those taught in all the schools and universities of the time. Quite simply he assumed that not only the earth but all bodies in the heavens were real, physical entities, hence subject, at least in principle, to all the pressures and forces exerted on earthly matter. In 1609, through his own telescope, Galileo had seen those heavenly bodies more clearly than anyone before him. He saw that the surface of the moon was rough and mountainous. Previously he had seen a "new star" among the supposedly fixed stars. The Aristotelians had assumed the heavens to be perfect, therefore immaterial and unchangeable; clearly, according to Galileo, they were wrong. But more than the assumption of perfection would have to go if Galileo's observations and mechanical experiments were correct. For example, Aristotle would have weight inhere in bodies and their speeds as freely falling bodies be proportional to their weight. Galileo argued that in motion, bodies fall at speeds determined not solely by their weight (or their shape) but by the resistance they encounter in the air—in short, that speed actually increases "the moment and force of the weight." And that speed, as well as that resistance, is measurable. In his words, "a material or corporal substance . . . has boundaries and shape, . . . relative to others it is great or small, . . . it is in this place or that, . . . it is moving or still, . . . it touches or does not touch another body"; and despite making every effort of imagination, Galileo claimed that he could not divorce that body from "these primary qualities." But its weight, taste, color, smell, these "secondary qualities" as he calls them, can be imagined away: "I hold that there exists nothing in external bodies . . . but size, shape, quantity, and motion" (*Il saggiatore,* 1624). In the universe constructed by Galileo and the other mechanical philosophers who were his contemporaries, heliocentricity was only one part of a larger conceptual whole. At its center lay the assumption that bodies and motion—both of which are capable of mathematical application and observation—are the objects on which the new science must now focus its attention.

Briefly stated, Galileo's presentation of this science, that body of learning he now explicitly offered to the educated laity as well as to the church, demanded the acceptance of a mélange of fundamental and new assumptions: The world around us consists of bodies subject to mechanical laws; these can be uncovered by the senses, by observation and experiment. The larger universe partakes in these operations (in ways Galileo would not have understood despite this assumption) because the earth is a body, like the planets, that moves in boundless space; and not least, if there are scriptural texts that state or imply otherwise, these must be understood as simply the use of commonplace metaphorical language intended for the benefit of "the vulgar," not as inherent contradictions between Scripture and natural philosophy. Galileo maintained that God's word could not ultimately contradict God's work. But Aristotle could be contradicted because, in

Galileo's words, "reasons persuade me—and Aristotle himself taught me to find peace of mind in that which I am persuaded by reason and not solely by the authority of the master . . . Philosophizing must be free." Once free, Galileo assures his readers, the science of mechanics and its branches may be quite useful "when it becomes necessary to build bridges or other structures over water, something occurring mainly in affairs of great importance."[11] When appealing to his audience Galileo also sought to persuade them of the usefulness of the new science.

The Culture of the Few and the Many

In choosing to argue for the utility of the new science Galileo sought an alliance between science and the material interests of the educated laity. In effect he endeavored to render science into an inherent part of a new and elite culture that had developed in the course of the sixteenth century, a culture increasingly distinct from popular culture and even hostile to it. Historians of early modern Europe have come to see the creation of this division between high and low culture as one of the central transformations occurring in that period. What is now also clear is that science became entwined in that transformation.[12]

During the sixteenth century three developments worked together to transform the culture of the people, and incidentally to provoke the reaction to it that permitted Galileo to make his appeal. The printing press and the Reformation promoted literacy, while the commercial revolution gave previously unknown men (and a few women) access to the press and publication because what they had to say, whether about religion, medicine, or mechanics, would sell in a world of increasing literacy, prosperity, and economic opportunity. The sixteenth century witnessed an expansion in the marketplace of Europe that, however, also coincided with constant inflationary pressures.[13] Put quite simply, the creation of an articulate popular culture, one that was occasionally heretical and hostile to the traditional magistrates, occurred simultaneously with a widening gap between rich and poor. All social and economic evidence we have from the period immediately prior to Galileo's confrontation with the church points to an increase in poverty for the majority in most parts of Europe, combined with an increasing prosperity for many aristocratic and mercantile elites, particularly for those who could use their land or capital to take advantage of the new inflationary or market pressures.[14]

In addition the Protestant Reformation, quite apart from its obvious appeal to the magistrates of the cities, and indeed to the heads of the new nation-states, offered ordinary people a vision, often millenarian in nature, of a better future order here on earth. This popular millenarianism, when combined with distinctively Protestant doctrines such as predestination and the priesthood of all believers, gave ordinary people a disciplined path by which they might achieve that better future order. We will find that millenarian vision an important rationale for the acceptance of the new science in Protestant countries; indeed the English philosopher Francis Bacon (1561–1626), a precise contemporary of Galileo, offered the new science as one of the avenues by which that millenarian reformation might be achieved (see pp. 33–35). But he did so in language that specifically repudiated any association between millenarianism and the culture of the people, or between science and the contemporary opponents of church and state.

In consequence of these earlier social, intellectual, and economic developments a dramatic change occurs in the traditional relationship between the culture of the few and the many in the course of the seventeenth century. As one historian has put it, "The elite, once having patronized popular culture, become increasingly suspicious and hostile to it. . . . The people and their culture had become dangerous."[15] The result is a gradual and widening separation between the culture of the many and the few. The elite, far from wishing any longer to foster popular culture, seek now to control and redirect it. It is within this context that we need to examine the specific case that Galileo makes for why the new science should be accepted and by whom.

In 1616 the theologians of the Congregatio Sanctae Inquisitionis condemned the proposition: *Sol est centrum mundi, et omnio immobile motu locali"* (The sun is the center of the universe . . .). They promulgated that decree to all the offices of the Inquisition in the world and also placed Copernicus's book on the revolution of the heavenly orbs on the Index of Forbidden Books. They acted in response to a letter they had had from a Florentine Dominican wherein he complained that the "Galileisti," that is, a group of Galileo's more aggressive followers, openly taught that the earth was moving. Prior to the sending of that letter, Galileo's clerical and Aristotelian enemies had attacked him from the pulpits of the city; indeed these enemies had formed a secret group that had as its expressed purpose the discrediting of Galileo and his doctrines.

By the time of this decree of 1616, Galileo was a figure of international reputation. As early as 1604 his lectures in Padua had attracted over one thousand listeners, and in 1610 he published a highly successful and easily read account of the new *supernova* that had appeared a few years earlier. When he became court mathematician to the Grand Duke of Tuscany, who resided in Florence, Galileo continued his aggressive search for converts to the new science, not the least of them the grand duke himself, whom Galileo assiduously wooed. In so doing he could be seen as challenging the monopoly on scientific education enjoyed by the clerical teachers of the local universities, many of whom may have learned nothing about astronomy since mastering Aristotle's *De caelo*. Their competence and hence their status within the academic community were directly challenged by the new science, and not surprisingly they spearheaded the attack on Galileo and used Scripture as their immediate weapon. As one Florentine Aristotelian put it when he used Aristotle's physics against Galileo and tied it to the preservation of a literal reading of the Scriptures: "All theologians without a single exception say that when Scripture can be understood according to the literal sense it must never be interpreted in any other way."[16] A similar warning was issued to Galileo by a cardinal of the church who stated in 1612 that Copernican doctrine could be maintained only if it was assumed that the Bible naively speaks about the earth's immovability "in accordance with the language of ordinary people." Such an assumption should be made only with extreme caution. Yet it was precisely that assumption which Galileo insisted upon making and which he openly proclaimed.

In 1615, while defending his own and Copernican ideas of the universe, Galileo insisted that "the mobility of the earth" is "a proposition far beyond the comprehension of the common people."[17] Galileo knew perfectly well, indeed was eventually told personally by the pope, that his scientific learning was still esteemed despite the decree of 1616, that he could always hold the

Copernican idea as a hypothesis. But he believed in more than that. Throughout his career, both before and after the condemnation of 1616, Galileo believed himself in possession of a special knowledge. He insisted that the new science was a discourse separate from the language of ordinary people and that the mechanical philosophy described the natural world better than any alternative explanation. With these assumptions he could hold to the truth of Copernican doctrine as well as to the laws of the new mechanics. He could also have the confidence and audacity to present both to the educated elite for their support. He may even have imagined that he possessed the power to influence the church at the very highest level of its hierarchy, in a circle to which he had frequent and long-standing access.

As a result of his self-confidence Galileo imagined as late as 1632 that it would be possible to have that earlier decree overturned. Indeed he castigated those clergy "who would preach the damnability and heresy of the new doctrine [that is, of the Copernican doctrine] from their very pulpits with unwonted confidence, thus doing impious and inconsiderate injury not only to that doctrine and its followers but to all mathematics and mathematicians in general."[18] The new science was unsuited to pulpit discussion—that is, to what Galileo and at least some of his closest friends and supporters had come to understand as the proper content of popular religiosity. One of Galileo's friends, Giovanni Ciampoli, explained to him the gulf he saw between their mutual scientific learning and what was now fit for popular dissemination:

> I have spoken to no one yet who did not judge it a great irrelevance for preachers to want to enter their pulpits and discuss such lofty and professional subjects among women and ordinary folk, where there exists such a small number of well-informed people.[19]

It must be remembered that at this time a Florentine Dominican had publicly attacked Galileo and his followers, indeed all "mathematicians," as being identical with astrologers. He had therefore attempted to slur the new science by associating it with magic and naturalism, beliefs still commonly found in folk culture. Throughout the seventeenth century the new science would seek to distance itself from those beliefs, indeed in some instances to wage war against them. In Galileo's defense of himself and his science we can see the first stage of that century-long struggle.

Galileo openly displayed his distance from "women and ordinary folk" in his various published defenses of the new astronomy against the strictures of the church. He argued that there are now two professional elites, mathematicians and theologians, and both have the obligation to take great care in what they say to the people. In other words both professional elites, mathematicians and theologians, must patronize the masses. The theologians, Galileo says, have long held that the Bible is full of passages "set down . . . by the sacred scribes in order to accommodate them to the capacities of the common people, who are rude and un-learned."[20] These passages are capable of a deeper meaning, one that it has always been the responsibility of theologians to discover.[21] At this point Galileo allies the new science with this exegetical tradition of maintaining an esoteric learning separate from, and unsuited for, the masses. He argues, "Hence even if the stability of heaven and the motion of the earth should be more than certain in the minds of the wise, it would still be necessary to assert the contrary for the preservation of belief among the all-too-numerous vulgar." The issue here, as Galileo presents it, is the danger of popular heresy: "The shallow minds of the common people" must be protected from the truth about the universe lest they "should become confused, obstinate, and contumacious in yielding assent to the principle articles that are absolutely matters of faith."[22]

As Galileo presented it, the new science constituted a body of knowledge fit for learned discourse that did not contradict the deeper meaning of Scripture. It was the "task of wise expositors to seek out the true senses of scriptural texts," leaving the commonplace language of the Bible intact for "the all-too-numerous vulgar." And in matters of that science Galileo is aggressive in his insistence that "in discussions of physical problems we ought to begin not from the authority of scriptural passages, but from sense-experience and necessary demonstrations."[23] God has given men their senses and reason, and we are obliged to use these "in physical matters." Galileo makes it absolutely clear that he and his followers have done a better job of using their senses and their reason than have their Aristotelian opponents. In arguing as he did that "wise expositors" must look beyond the literal meaning of Scripture, Galileo unwisely put himself in disagreement with the decree of the Council of Trent (1546), which prohibited any attempt "to twist the sense of Holy Scripture against the meaning which has been and is being held by our Holy Mother Church."[24] That decree had been issued in direct response to the Protestant Reformation and the multitude of new biblical interpretations that had been offered by learned Protestant theologians and also by the numerous unlearned Protestant sects that had sprung up all over Europe.

The Aristotelian opponents of Galileo, such as the Florentine Lodovico Colombo, managed to have their version of how the universe works accepted by the church's hierarchy because they were able to embrace the teachings of Trent at a time when the church feared any new voices, even those that attempted to restrict their learning to the discourse of the educated. In 1632 Galileo was put on trial by the Inquisition, and the following year he was condemned to house arrest. From that moment, all that he published had to be smuggled out of Italy to the freer presses of the Dutch cities. He had lost his struggle with his clerical opponents, and other Christian supporters of scientific learning saw his defeat precisely in those terms. The English poet John Milton visited him in Italy in 1638 and wrote, "There it was that I found and visited the famous Galileo, grown old, a prisoner to the Inquisition for thinking in Astronomy otherwise than the Franciscan and Dominican licensers of thought."[25]

THE EFFECT OF GALILEO'S CONDEMNATION

The writings of Galileo, and his subsequent trial and condemnation, moved the new science into the forefront of learned discourse throughout Europe. Anyone attracted by the ideas of Copernicus, if living in Catholic as opposed to Protestant Europe, now had to think very carefully about how to announce that support. In France, for instance, the clerical opponents of papal intervention in the affairs of the French church saw in Copernicanism a new weapon in their struggle; the Jesuits, with their strongly ultramontane (propapal) conception of religious authority, sided with the Inquisition's condemnation.[26] In Protestant countries, on the other hand, support for Copernicanism could now be construed as antipapal and hostile to the power of the Catholic clergy. What a splendid incentive for its adoption. This ideological linkage was to prove critical in creating the alliance between Protestantism and the new science.

Empirical science continued to be practiced in Italy after the public condemnation of Galileo. Yet the major philosophical innovations were now to occur elsewhere. After his condemnation, science in the seventeenth century became an increasingly Protestant and therefore northern and western European phenomenon. Much controversy has been generated among historians trying to explain that linkage, but the relationship can be sustained if we emphasize two points. The first is the ideological link, so attractive to Protestants, between opposition to the powers of the Roman church and its clergy and support for Copernicanism —and it should be borne in mind that those powers were frequently justified philosophically by the use of Aristotelian arguments. The second point concerns the dissemination of scientific knowledge. It is a truism that the enterprise of science depends on the communication of new knowledge; in early modern Europe that meant on the printing of scientific books. After the condemnation of Galileo, books at the vanguard of the new science—that is, those that advocated the mechanical philosophy and heliocentricity—had to be published where the Inquisition had no authority. In practice that meant in Protestant Europe: in the German cities, in England, and most especially in the Dutch Republic, which had only just won its independence from Spain.

The history of the printing press is inextricably tied to the history of the Scientific Revolution from Galileo to Newton, and the history of both became increasingly Protestant in the course of the seventeenth century.[27] As we shall see in Chapter 2, this does not devalue the importance of the ideological synthesis achieved by French science, that is, by Cartesianism. It does, however, offer one explanation for a complex process that permitted the new science by the early eighteenth century to achieve an unprecedented degree of cultural integration, first in England and then in the Dutch Republic. In both places cultural leadership and political authority were overwhelmingly in Protestant hands. By the latter part of the seventeenth century the new science had become institutionalized in England; it had also become an essential part of the way in which a few educated and influential men, and some women, understood the world around them. In the early eighteenth century that English science in its Newtonian form was first accepted on the Continent in the Dutch Republic and disseminated there by Dutch scientists as well as by the French language press based in the Dutch cities. Many historical consequences can be traced back, in some ultimate sense, to the "victory" over Galileo of a few Aristotelian professors, Florentine clergymen, and the bureaucracy of the Roman Inquisition.

Galileo's confrontation with the Roman church subsequently revealed the enormous power of the church and the Counter Reformation to control Italian intellectual and political life, to censure that which it regarded as heresy. It also brought into open discussion the entire issue of what science should mean to the secular laity as well as to "the people." Both issues, the orthodoxy of science as well as its uses, would remain central, and controversial, in the history of the integration of science in western culture.

THE DANGER OF NATURALISM AND ENTHUSIASM

In the late sixteenth century the Italian reformer Giordano Bruno, a Dominican by training as well as a humanist of sorts, attempted to use Copernicanism as the centerpiece of a new religious philosophy, one intended by him as an alternative to the positions staked out by what he regarded as the rigid dogmatism of both the Reformation and the Counter Reformation. He traveled throughout the courts and universities of Europe seeking support for what was in effect his new religion. At the universities in Germany and England, at the French court, Bruno proclaimed his new philosophy in opposition to what he called "the moral blindness of the new order," and in particular he opposed it to the growing power of the alliance between the papacy and the Spanish monarchy.[28] He too sought to capture the attention of the educated and secular elite, and he did not hesitate to attack what he called the "pedants" of the universities.

In the process Bruno attempted a synthesis of Copernican heliocentricity with a vision of nature that was both ancient and pagan, and extremely heretical. In Bruno's hands the sun is at the center of the universe, and the motion of the earth renders it into a living being; the universe is infinite and its planets inhabited. Indeed all of Nature is alive for Bruno; that is a mystical, almost magical truth, a mystical "furor" that, he claimed, renders the dogmatism of the pedants and the churches largely irrelevant. Suddenly the new science has been brought into alliance with forms of religious belief and magical practice long regarded by the church, and sixteenth century Protestants, as extremely dangerous. We shall call these distinct beliefs by the terms that seventeenth-century contemporaries would have understood—*naturalism* and *enthusiasm*—and we shall see them reappearing at other moments in the social history of the new science.

Naturalism in its heretical and Brunian form preached that nature is in effect alive; it could be worshiped in itself. On the one hand it might be the object of increasing scientific inquiry, while on the other its hidden workings, its force and power, could be embraced by reforming philosophers intent on destroying the power of the organized churches. The symbolic power of nature could be made to rival the ritual and dogma presented with such grandeur by the clergy. By extension, might not science itself be proclaimed as the religion of those who would worship only the world around them? That was a usage for the new science that all of its major promoters feared.

By having attacked the commonplace Aristotelianism of the schools, as well as the right of the church to censor books, the new scientists had themselves opened the door, however unwittingly, to reformers such as Bruno who preached naturalism and did so in a way that contemporaries described as leading to "enthusiasm." Enthusiastic religion had long been associated with the undisciplined religiosity of "the people." It relied on a mystical belief in the power of ordinary men and women to communicate with the Deity; it presumed that their wisdom equaled that of the learned. The religious upheavals of the sixteenth century had renewed the fear of popular enthusiasm; and by the middle of that century there were small sects throughout Europe, such as the Anabaptists,* who were indeed, from the perspective of both Protestant and Catholic clergy and magistrates, enthusiasts.[29] Bruno sought to merge naturalism, of a learned variety, with Copernicanism and to render both into the foundations

*See glossary of terms.

of a new religion that, if preached to the masses, had the potential, it was believed, to kindle their "enthusiasm." That volatile mixture brought Bruno to his death at the hands of the Roman Inquisition.

While Galileo may have known little about Bruno, some scholars have argued that the fear of giving support to Bruno's mystical and dangerous form of religiosity may have lurked in the minds of Galileo's accusers.[30] After the execution of Bruno, the Inquisition burned every copy of his books that it could find; they remain, to this day, extremely rare. But naturalism had many roots and many advocates in the early modern period, and it would return again and again to haunt Christian natural philosophers intent on winning the educated to their ideas but terrified of what the new science might be used to support. As we shall see in our discussion of eighteenth-century religiosity, they had good reason to be worried (see pp. 120–121).

In Italy the prisons of the Inquisition housed another anti-Aristotelian reformer, Thomas Campanella (1568–1639). He was "detained" in 1597 for attacking the doctrines of Aristotle from a naturalistic perspective that derived from the contemporary revival of those doctrines among Italian humanists. Campanella combined heretical philosophizing with rebellious political opposition to Spanish influence in parts of the Italian peninsula, and in 1599 he was tortured and sentenced to life imprisonment for his part in a conspiracy to overthrow Spanish rule in Calabria. From his cell he wrote a passionate *Defense of Galileo* in 1616, which was published in 1622 in Frankfurt.[31] The text denounced as "mad" theologians "who believe that Aristotle constructed the true system of the heavens"; and it claimed that Copernicus had in fact "turned back to the ancient doctrines of the Pythagoreans, where better explanations of the phenomena are presented. Following him, Galileo discovered new planets and new systems, and unknown changes in heaven."[32] Campanella also believed that the language of the Bible had been intended for the "vulgar" and that it behooved the learned to embrace the new science of Galileo as the foundation for all true philosophy. Campanella knew the writings of Bruno, and indeed his own mystical religiosity closely resembles the naturalism that Bruno offered as an alternative to the Counter Reformation led by Spain.

But Campanella's faith in the power of natural philosophy to transform ignorance into wisdom led him to part company with

Galileo's insistence that the new science was unfit for "the vulgar." In 1620 Campanella sought to win the favor of his captors and became an apologist for the Spanish king.[33] In his *Monarchia di Spagna* (translated into English in 1654 as *Discourse Touching the Spanish Monarchy*), Campanella sets forth a new view of the connection between natural philosophy, political and religious authority, and the common people—a view very different from Galileo's and remarkably "progressive" in that it, quite unintentionally, anticipates developments that will occur late in the eighteenth century (see pp. 203–204).

As James R. Jacob has argued, Campanella agreed with Galileo that ordinary men and women could be "contumacious" (to use Galileo's word) toward authority:

> As a reformed revolutionary, Campanella would have known whereof he wrote. But here he departs from Galileo. The latter had said that natural philosophy should be kept from "the vulgar" because it was likely to make them more unruly. Campanella, on the other hand, argues that the study of nature should be encouraged among the people in order to promote obedience to authority. Campanella was convinced that natural philosophy, rather than causing people to question authority, would absorb their interest and hence turn their attentions away from politics and religious dissent. Science would also channel human energies into productive enterprise, energies that might otherwise go to fuel popular disturbances.

By means of scientific education, Campanella says, "people's minds will be diverted from creating . . . any trouble, and will be incited to bend their studies that way which may be useful to the king."[34] Were popular scientific education to be undertaken it would serve two purposes, the one technical and in an attenuated sense industrial, the other political. By the late eighteenth century the promoters of industrialization and scientific education in England and the Continent, had they known that tract by Campanella, would have agreed.

THE SOCIAL UTILITY OF SCIENCE

Campanella may have been the first in Italy, or anywhere in Europe for that matter, to make this striking argument for the integration of science into the thinking of all social groups, but he was by no means the last. As Carlo Ginzburg has shown, later in the century, after Galileo's trial and condemnation, the Italian thinkers Sforza Pallavicino and Virgillio Malvezzi made a very similar argument for the social utility of science. It is right, they argued, that ordinary people—artisans and peasants—should be kept out of politics, which is tricky and unpredictable. Politics presupposes a secret wisdom known only to princes. But science can be made available to everyone because nature is everywhere the same: its operations, unlike those of politics, are regular and predictable. Thus scientific inquiry can be safely encouraged among the populace, and to the extent that ordinary people commit themselves to the study of nature, politics can become what it ought to be, a monopoly of the elite, with a consequent reduction of disorder produced by popular rebellion.[35]

This argument for the widespread practice of scientific inquiry rests on its supposed social utility in maintaining traditional authority. It is first enunciated, although never put into practice, in Counter Reformation Italy at least partly in response to the church's condemnation of the new astronomy and particularly of Galileo's defenses and demonstrations of it. It was to become a powerful argument, one that we shall later hear from English apologists for the Royal Society; indeed a version of it appears in Thomas Sprat's official *History* (1667) of the Society. In that Protestant context, where clergy and scientists could be allied in a common enterprise intended to support the Church of England, the argument gained acceptance. Eventually, in the eighteenth century, the power promised by science became one of the most important justifications for the promotion of scientific inquiry. Yet before it could be accepted science had to be rendered safe.

The English version of the social utility argument was probably developed quite independently from the Italian ones. What is important about the argument is that it appears in a variety of contexts, but always with the same intention. Science can increase the wealth and power (both social and military) of existing elites. It can be a force for social stability, not social reform, and its purpose is to increase the prosperity and wealth of the state. In every period where we find that argument put forward with particular force we will also find radical thinkers who oppose it, who would have science directly serve the people, to the benefit of all humankind.

These radical voices will be heard during the English Revolution, and again throughout Europe in the late eighteenth century (see p. 204). By that time we once again find Italian reformers in touch with the larger European phenomenon we describe as the Enlightenment who will participate in that reforming movement (see pp. 206–208). But in the period after Galileo's condemnation, advocates of the new science had to smuggle their writings out of Italy; and from the perspective of the general history of Europe, science found a safe home north of the Alps—in France, the Dutch Republic, and most especially England.

In his writings Galileo laid the foundation for the science of mechanics, the study of the forces that act on bodies in motion. Propagators of mechanics throughout the seventeenth and eighteenth centuries who sought to render their science practical and applicable frequently acknowledged its Galilean origins. In addition Galileo proclaimed the centrality of science in the new culture of the literate classes. In effect he had argued as a civic humanist that learning should serve the polity, but he focused that argument quite precisely on mechanical knowledge based on a new philosophy of nature. He appealed for support to the merchants and aristocrats of the Italian city-states, but it never occurred to him that civic humanism could be placed in an even grander secular alliance, in the service of monarchs and entire national states.

THE BACONIAN VISION

That grander vision, not surprisingly, came from a servant of one of the national states, the Lord Chancellor of England, Francis Bacon (1561–1626). In the first instance Bacon sought to render monarchical government increasingly effective, and he wrote for his king, James I, whom he wanted to cajole into giving his support to "the advancement of learning." But in his definition of learning Bacon gave special place to science, which he characterized primarily as the widest possible accumulation of knowledge about nature. Indeed in the first instance he envisaged a vast program for the collection of information on every aspect of nature; his empiricism knew no bounds. Yet Bacon possessed a very precise sense of the utility of knowledge and saw, with remarkable perspicuity, that the mechanical arts could make an unprecedented contribution "to the endowment and benefit of man's life."[36] In this respect Bacon saw more clearly than any of his contemporaries the extraordinary advances that had already been made by mechanical artisans in shipbuilding, navigation, ballistics, printing, and water engineering.[37] He also knew the disdain in which the educated and titled held such unlettered men. He chided, "It is esteemed a kind of dishonor unto learning to descend to inquiry or meditation upon matters mechanical, except they be such as may be thought secrets, rarities, and special subtilties." He attacked that "supercilious arrogancy,"[38] so much a part of the aristocratic culture of his time, and in its place he offered a new vision of the truly cultured and educated man.

Such a man—and Bacon was very precise in giving a gender identity to scientific activities as he sought to assure the aristocracy and gentry that science is truly "masculine"—repudiated the "degenerate learning [that] did chiefly reign amongst the schoolmen."[39] In *The Advancement of Learning* (1605) Bacon directly assaulted the old clergy of the Roman church, "their wits being shut up in the cells of a few authors (chiefly Aristotle their dictator) as their persons were shut up in the cells of monasteries and colleges." In so doing he allied scientific learning with Protestant culture of the kind that had been institutionalized after the Henrican Reformation: Erastian, in that it favored the domination of bishops and ministers by king and local aristocracy, and national, in that it eschewed the sectarian divisions that were commonplace in Continental Protestantism and opted for a single national Anglican church. Such Protestant gentlemen as Bacon would create should cultivate the sciences and observe the activities of the mechanical artisans so as to achieve a natural philosophy that would be progressive, one capable of "perpetual renovation." What better way to preserve governments, he asked, than "to reduce them, *ad Principia* [to first principles], [to] a rule in religion and nature, as well as in civil administration?"[40] The Protestant state would flourish, Bacon believed, under the aegis of a wise central administration led by a wise king, a unified church, and not least, a common enterprise of scientific endeavor.

Not all the learned men of science would pursue the same activities. Bacon left ample room for a variety of scientific activities, "some to be pioneers and some smiths, some to dig and some to refine and hammer," some to be "speculative, and [others] operative." For Bacon, as for all his early modern followers, this division of labor was two sides of the same coin. The modern distinction between "pure" and "applied" science is a nineteenth-century invention; it was simply not understood in this earlier period. Consequently we must follow Bacon's language in our discussion of seventeenth- and eighteenth-century science. There were no scientific amateurs in contradistinction to the professionals. Instead, as Bacon would explain in his posthumous utopian tract, *The New Atlantis* (1627), there would be many different workers in the vineyard of science. There would be "the merchants of light," those who would use "the books, and abstracts, and patterns of experiments" to propagate science in the market place. Contemporaries inspired by this utopian vision interpreted it as the call for a vast, European-wide program for the dissemination of science, one unprecedented as an ideal before or since. In his utopian paradise Bacon would also have "Lamps," those who would take care "to direct new experiments, of a higher light, more penetrating into nature than the former," that is, those who would draw out of science "things of use and practice for man's life."[41]

In his advocacy of the validity and usefulness of the new science, Bacon repudiated the secretiveness and exclusivity of the magician; he also urged the rejection "of fables and popular errors."[42] His science enshrined the work ethic, "the laborious and sober inquiry of truth," as the correct method of inquiry in opposition to "the high and vaporous imaginations" found in natural magic, in superstition, and in Aristotelian emphasis on "sympathies and antipathies, and hidden proprieties." Eloquently he allied this sober, mechanically oriented science with the Protestant Reformation:

> We see before our eyes, that in the age of ourselves and our fathers, when it pleased God to call the Church of Rome to account for their degenerate manners and ceremonies, and sundry doctrines obnox-

ious and framed to uphold the same abuses; at one and the same time it was ordained by the Divine Providence, that there should attend withal a renovation and new spring of all other knowledge.[43]

Bacon saw the renovation of the sciences as the work of divine providence, and as a true English Protestant he possessed a very precise sense of the role of providence in history. This renovation was part of a larger scheme, a vast unfolding, a "great instauration" of learning that would precede the end of the world and liberate human beings from the effects of their original fall from grace.[44] This millenarian impulse, so much part of English Protestantism and, particularly after Bacon's death, English Puritanism, must be reckoned as one of the main motivations for the cultivation of scientific inquiry in seventeenth-century England. The Baconian vision became inextricably bound up with it. The millenarianism of Isaac Newton and some of his followers later in the century can also be related to this background (see pp. 84–87). Similarly English scientists' militant promotion of science as one of the foundations, both practical and ideological, of state power owed much to their unique and compelling search for the millenarian transcendence of historical time. It should be emphasized, however, that the millenarianism of Bacon and his followers always located control and leadership in the millenarian paradise firmly in elite hands.[45]

Yet an emphasis on Bacon's millenarianism inevitably entails a recognition that there were profoundly mystical elements in his thought. He would steal the vision of the magicians by unlocking nature's secrets, yet he would discard their secretive methods. Bacon believed that in the ancient myths and fables lay a hidden wisdom, and in this search to recover and vastly augment that wisdom he resembles Bruno as well as various hermetic thinkers of the sixteenth century, not least the Swiss medical reformer Paracelsus (1494–1541). Paracelsus used the magical and Platonic tradition, which emphasized the correspondences between the human body and the heavens to legitimate an empirical and experimental approach to the study of disease, and his program was blatantly anti-Aristotelian. Bacon condemned the magical elements in the thought of Paracelsus, but he had to admit that the latter's natural history was extremely useful.[46] The magical tradition in early modern Europe could, and indeed at moments did, promote scientific inquiry. Magic could promote the search for alternative natural philosophies to that of Aristotle; its literature on alchemy and astrology could also promise an unfolding of nature's secrets, and its systematic exploration could promote empiricism. At their core the magical arts promised to reveal a single, unifying philosophy of nature. For that reason Bruno could be a Copernican of sorts; Bacon could attempt to steal the zeal of the magician while excluding him from the company of the experimenters;[47] Newton could practise alchemy throughout his career; and the German scientist and mathematician Gottfried Wilhelm Leibniz (1646–1716) could dabble in astrology. Yet in its adoption of the mechanical philosophy and in its repudiation of popular beliefs, the new science of the early modern period ultimately rendered magic irrelevant to the needs and interests of the educated elite.

The profoundly humanistic quality of the Baconian vision made it attractive to scientifically minded social reformers as well as to promoters of the new science throughout the seventeenth and eighteenth centuries, if not well beyond. The belief that science made useful could bring "the relief of man's estate" has never been surpassed as a rationale for natural philosophical inquiry and general scientific education. As early as the 1620s Bacon's ideas were known on the Continent, particularly in select philosophical circles in Paris.[48] There his emphasis on the systematic collection of data, on an easily applied empiricism, appealed to botanists and collectors at the newly established Jardin des Plantes. Similarly his vision that science promised to lighten the human burden, to provide domination over nature, appealed to reforming German Protestants of the same period who wished, like Bacon, to include natural knowledge in the millenarian reforms promised in the coming age.

In a self-conscious bid for the widest possible readership among the literate in England, Bacon published most of his important works in English and not in Latin. Yet they were quickly translated into Latin and published in Continental editions generally emanating from Amsterdam. His writings were known as early as 1620 to the important Dutch mechanical philosopher Isaac Beeckman (see pp. 52–54). During the 1640s Bacon's ideas were discussed at the University of Leiden. The Netherlands in general and Leiden in particular were the most important centers for natural philosophical education in seventeenth-century Europe. At the end of the century the most advanced medical teacher and practitioner of his time, Herman Boerhaave (1668–1738), a professor at Leiden, could barely contain his rhetorical enthusiasm for the promise of medical advancement offered to those who would heed Bacon's call to experience nature for themselves.[49] Perhaps even more fascinating in terms of the general spread of the Baconian vision, we can find his utopian and humanistic writings about science, among other topics, translated into Dutch, also in the 1640s and 1650s.

The New Atlantis, Bacon's vision of an island paradise dedicated to peace and scientific progress, was first published in English after his death. It was immensely popular in England, and in 1656 a pocket-size edition appeared in The Netherlands, in a Dutch translation that purposefully sought to be clear and simple. It would seem that Bacon may also have appealed to the literate in one of the most heavily mercantile and urbanized areas of Europe.[50] In the spirit of Bacon's search for powerful patrons to promote the work of science, that translation was dedicated to a prince, in this case Frederic Henry, Prince of Orange. Given the diffusion of millenarian ideas among Continental Protestants, we may postulate that Baconian ideas about a future paradise based on science also evoked interest born of millenarian fervor.

By the eighteenth century reformers, no longer convinced that human progress required any disruption of historical time, easily ignored Bacon's millenarian side and concentrated on his call for a scientific empiricism intended for the relief of the human condition. The French encyclopedists, led by Diderot and d'Alembert (see pp. 202–205), invoked the memory and ideals of Francis Bacon in the first great *Encyclopedié* (1751), which sought to make all knowledge unified and accessible. In the 1790s we find Bacon's *Novum Organum* (1620) translated into German at a time when German scientific societies were proliferating and the old professoriate was being challenged by newer men profoundly interested in the practical application of the sciences to the problems of society and industry.[51] In the 1830s one branch of the leadership of the British Association for the Advancement of Science, an organization committed to the practical and industrial application of science, also invoked the memory and vision of Bacon. By that time, however, Bacon's vision of a science based on the collection of facts could be opposed to a more theoretical and professionalized vision of science, one whose supporters wanted the enterprise of science to be dominated by "great minds" and by the search for general laws.[52] This alternative to the Baconian vision they, and we, could describe as Newtonian, and even by the early eighteenth century in England it had come to replace Baconianism as the dominant scientific ideology.

SCIENCE IN ELITE CULTURE

The early seventeenth-century prophets of the new science, however much they might differ on its religious meaning or on its dissemination outside a narrow elite, recognized the power that lay within the grasp of those who would support their enterprise. They wanted to divorce science from popular beliefs while simultaneously recognizing that mechanical knowledge of a practical kind occasionally resided among artisans. Yet a philosophically mechanical understanding of nature required a higher learning not intended for the people, as Galileo was quick to point out. It arose in part out of experimentation with mechanical devices of every kind and also depended on the observation of actual bodies in motion in the terrestrial and celestial worlds. It required for its success the abandonment of Aristotle or, more precisely, a repudiation of his contemporary teachers, who, if we are to believe their critics, sat in their universities unaware of the world of moving bodies around them. But in the cities of late Renaissance Italy, in the mercantile towns of the Dutch Republic, in the courts of the powerful new monarchs, followers might be found who would apply their literacy toward mastering a new body of learning and, even more important, a new philosophy of nature.

The promise of mastery over nature must have been overwhelming to those who could believe in it as a serious alternative to the mastery that nature had held for so long over the well-being of all humankind. These first appeals to follow the new science as an alternative to traditional and popular beliefs about nature were somewhat vague as to what might be accomplished. In the hands of the naturalists they were also ominous for the traditional order and authority over society deemed necessary by both lay and clerical elites. The scientific ideology formulated by the next generation of mechanists aimed to rectify both those errors. As formulated by Descartes in the 1630s and by his later followers, the new mechanical philosophy offered precise applications as evidence of its validity, and it held out the promise of a complete philosophy that would reinforce social order and religious orthodoxy. Throughout seventeenth-century Europe natural philosophy gave expression to values in the "world politick," just as it attempted to describe mechanically the "world natural." At every turn, that linkage ensured its integration into the larger culture and made its ideological formulations immediately and directly relevant to those who held, or sought to hold, power in society and government.

NOTES

1. Galileo Galilei, *Two Chief World Systems,* trans. S. Drake (Berkeley: University of California Press, 1967), p. 207.
2. Ibid.
3. R. Hooykaas, "Rheticus' Last Treatise on Holy Scripture and the Motion of the Earth," *Journal for the History of Astronomy,* vol. 15 (1984), pp. 77–80. This phrase has been interpreted as an acceptance of heliocentricity as mere hypothesis, but that is not what the sentence says.
4. See William A. Wallace, ed., *Galileo's Early Notebooks: The Physical Question* (Notre Dame, Ind.: University of Notre Dame Press, 1977). For a balanced assessment of science at one of the major Protestant universities of the century, see John Gascoigne, "The Universities and the Scientific Revolution: The Case of Newton and Restoration Cambridge," *History of Science,* vol. 23, no. 62 (1985), pp. 391–434.
5. For the career of Bruno, see the classic study, Frances Yates, *Giordano Bruno and the Hermetic Tradition* (London: Routledge and Kegan Paul, 1964).
6. This point is made in the extremely useful essay by Olaf Pedersen, "Galileo and the Council of Trent: The Galileo Affair Revisited," *Journal for the History of Astronomy,* vol. 14, no. 39 (1983), p. 24.
7. Vincenzo Ferrone, "Galileo tra Paolo Sarpi e Federico Cesi," in *Novita Celesti e Crisi del Sapere* (Atti'del Connegno Internazionale di Studi Galileiani, 1983), pp. 239–253. Cf. A. Banfi, *Vita di Galileo Galilei* (Milan: Feltinelli, 1962).

8. For this point, see Paul W. Knoll, "The Arts Faculty at the University of Cracow at the End of the Fifteenth Century," in Robert S. Westman, ed., *The Copernican Achievement* (Berkeley: University of California Press, 1975), pp. 137–156.

9. Nancy Siraisi, *Taddeo Alderotti and His Pupils: Two Generations of Italian Medical Learning* (Princeton: Princeton University Press, 1981).

10. Quoted in Robert S. Westman, "Three Responses to the Copernican Theory: Johannes Praetorius, Tycho Brahe, and Michael Maestlin," in Robert S. Westman, ed., *The Copernican Achievement* (Berkeley: University of California Press, 1975), pp. 289–303. For resentment against Aristotle at Edinburgh in the late 1630s see E. Forbes, "Philosophy and Science Teaching in the Seventeenth Century," in Gordon Donaldson, ed., *Four Centuries: Edinburgh University Life 1583–1983* (Edinburgh: University of Edinburgh Press, 1983), p. 30.

11. Stillman Drake, *Cause, Experiment and Science: A Galilean Dialogue Incorporating a New English Translation of Galileo's "Bodies that Stay atop Water or Move in It"* (Chicago: University of Chicago Press, 1981), p. 21.

12. Natalie Davis, *Society and Culture in Early Modern France* (Stanford: University of California Press, 1975), pp. 189–267; and Peter Burke, *Popular Culture in Early Modern Europe* (New York: Harper & Row, 1978), pp. 245–281. See also Anne Jacobson Schutte, "Teaching Adults to Read in Sixteenth Century Venice: Giovanni Antonio Tagliente's *Libro Maistrevole*," *Sixteenth Century Journal,* vol. 17, no. 1 (1986), p. 3–16.

13. A. G. Keller, *A Theatre of Machines* (New York: Macmillan, 1964), pp. 2–5; Davis, *Society and Culture,* pp. 214–217, 221, 225; Paolo Rossi, *Philosophy, Technology and the Arts in the Early Modern Era* (New York: Harper & Row, 1970), pp. 1–62; and most important, Elizabeth Eisenstein, *The Printing Press as an Agent of Change,* 2 vols. (Cambridge: Cambridge University Press, 1978), vol. 1, pp. 303–378.

14. Peter Kriedte, *Peasants, Landlords and Merchant Capitalists,* English translation (Leamington Spa, U.K.: Berg Publishers, 1983), pp. 57–64; for Florence see p. 72.

15. The points made here first appear in James R. Jacob, " 'By an Orphean Charm': Science and the Two Cultures in Seventeenth-Century England," in Phyllis Mack and Margaret C. Jacob, eds., *Politics and Culture in Early Modern Europe: Essays in Honor of H. G. Koenigsberger* (Cambridge: Cambridge University Press, 1986), pp. 231–232.

16. As quoted in Pedersen, "Galileo and the Council of Trent," p. 6. I am indebted to this account, which adds considerable evidence to the argument first put forward in James R. Jacob and Margaret C. Jacob, "The Social Foundations of Modern Science: Historiographical Problems," paper presented at a joint session of the American Historical Association and the History of Science Society, Los Angeles, December 1981.

17. Galileo Galilei, *Letter to the Grand Duchess Christina,* in Stillman Drake, ed., *Discoveries and Opinions of Galileo* (Garden City, N.Y.: 1957), p. 169.
18. Ibid., p. 177.
19. Drake, *Discoveries and Opinions,* p. 161. Quoted by J. R. Jacob in " 'By an Orphean Charm,' " p. 239.
20. Galileo in *Drake, Discoveries and Opinions,* p. 181. The point is made in J. R. Jacob, " 'By an Orphean Charm,' " p. 239.
21. Ibid., pp. 181–182.
22. Ibid., p. 200 for both quotations.
23. As quoted in Pedersen, "Galileo and the Council of Trent," p. 18.
24. Ibid., p. 15.
25. Quoted in Stillman Drake, ed., *Galileo Galilei's Dialogue Concerning the Two Chief World Systems,* (Berkeley: University of California Press, 1967), p. xxv.
26. Georges Gusdorf, *La Revolution Galiléenne* (Paris: Payot, 1969), vol. 1, pp. 122–124.
27. See Eisenstein, *The Printing Press.*
28. Frances A. Yates, *Lull and Bruno: Collected Essays* (London and Boston; Routledge and Kegan Paul, 1982), in particular "The Religious Policy of Giordano Bruno," pp. 151–179.
29. George H. Williams, *The Radical Reformation* (Philadelphia: Westminster Press, 1962), pp. 321–322, 599, 609.
30. That interpretation is argued in E. A. Gosselin and L. S. Lerner, "Galileo and the Long Shadow of Bruno," *Archives internationales d'histoire des sciences,* vol. 25 (1975), pp. 222–246.
31. For that text, see William Dodge Gray and Harold U. Faulkner, eds., *Smith College Studies in History,* vol. 22, 1937, which contains Grant McColley, ed., *The Defense of Galileo.* Reprinted 1975 by Arno Press.
32. Ibid., pp. 22–23.
33. Yates, *Bruno,* pp. 360–367, 385–389. These paragraphs were first published in J. R. Jacob, " 'By an Orphean Charm' " (see note 15), pp. 239–241.
34. Quoted in Henry Stubbe, *Campanella Revived* (London, 1670), p. 3. Cf. James R. Jacob, *Henry Stubbe: Radical Protestantism and the Early Enlightenment* (Cambridge: Cambridge University Press, 1983), p. 86.
35. Carlo Ginzburg, "High and Low: The Theme of Forbidden Knowledge in the Sixteenth and Seventeenth Centuries," *Past and Present,* no. 73 (November 1976), pp. 28–41, especially p. 37. This paragraph was first published in J. R. Jacob, " 'By an Orphean Charm,' " p. 240.
36. Francis Bacon, *The Advancement of Learning,* ed. Arthur Johnston (Oxford: Clarendon Press, 1974), p. 71.
37. Edward Zilsel, "The Sociological Roots of Science," *American Journal of Sociology,* vol. 47 (1942), pp. 245–279. On the role of architectural

artisans, see Pamela O. Long, "The Contribution of Architectural Writers to a 'Scientific' Outlook in the Fifteenth and Sixteenth Centuries," *Journal of Medieval and Renaissance Studies,* vol. 15, no. 2 (1985), pp. 265–298. On practical mathematicians advocating experimentation, see J. A. Bennett, "The Mechanics' Philosophy and the Mechanical Philosophy," *History of Science,* vol. 24, no. 63 (1986), pp. 1–28.

38. Bacon, *The Advancement of Learning,* p. 70.
39. Ibid., p. 27. For gender identity in Bacon's thought, see Carolyn Merchant, *The Death of Nature: Women, Ecology and the Scientific Revolution* (San Francisco: Harper & Row, 1980). On the seventeenth century in general, see Susan Bordo, "The Cartesian Masculinization of Thought," *Signs,* vol. 11, no. 3 (1986), pp. 439–456.
40. Bacon, *The Advancement of Learning,* p. 70.
41. Francis Bacon, *The New Atlantis,* ed. Arthur Johnston (bound with *The Advancement of Learning* previously cited; Oxford: Clarendon Press, 1974), pp. 245–246.
42. Bacon, *The Advancement of Learning,* p. 69.
43. Ibid., p. 42.
44. On Bacon and the Apocalypse, see Katharine R. Firth, *The Apocalyptic Tradition in Reformation Britain, 1530–1645* (Oxford: Oxford University Press, 1979), pp. 204–207.
45. See J. R. Jacob, " 'By an Orphean Charm,' " pp. 241–245.
46. P. M. Rattansi, "The Social Interpretation of Science in the Seventeenth Century," in Peter Mathias, ed., *Science and Society, 1600–1900* (Cambridge: Cambridge University Press, 1972), pp. 12–18.
47. For a splendid exposition of this point see Paolo Rossi, *Francis Bacon: From Magic to Science* (London: Routledge and Kegan Paul, 1968), passim. See also Charles Webster, *From Paracelsus to Newton: Magic and the Making of Modern Science* (Cambridge: Cambridge University Press, 1982).
48. See Rio Howard, "Guy de La Brosse: Botanique et chimie au début de la revolution scientifique," *Revue d'histoire des sciences,* vol. 31 (1978), pp. 325–326.
49. Th. H. L. Scheurleer and G. H. P. Meyjes, eds., *Leiden University in the Seventeenth Century* (Leiden: Brill, 1975), p. 312; and see E. Kegel-Brinkgreve and A. M. Luygendijk-Elshout, eds., *Boerhaave's Orations* (Leiden: Brill and Leiden University Press, 1983), p. 177.
50. *Nieuwen Atlas, Ofte Beschrijvinge van het noytmeer gevonden Eylandt van Bensalem,* trans. J. Williaemson (Dordrecht, 1656). See also Franciscus Bacon, *De Proef-Stucken,* trans. Peter Boener, apothecary of Nijmegen—a translation of Bacon's moral and religious essays and his *Wisdom of the Ancients.* The copy at the University Library, Amsterdam, is from the library of Constantine Huygens. This is a very rare edition.
51. See *Neues Organon aus dem Lateinischen ubersetzt von George W. Bartoldy*

(Berlin, 1793). See also Steven Turner, "The Prussian Professoriate and the Research Imperative 1790–1840," in H. N. Jahnke and M. Otte, eds., *Epistemological and Social Problems of the Sciences in the Early Nineteenth Century* (Dordrecht: Reidel, 1981), pp. 116–118.

52. Jack Morrell and Arnold Thackray, *Gentlemen of Science: Early Years of the British Association for the Advancement of Science* (Oxford: Clarendon Press, 1981), pp. 267–273. See also Richard Yeo, "An Idol of the Market-Place: Baconianism in Nineteenth Century Britain," *History of Science*, vol. 23, no. 61 (1985), pp. 251–298.

CHAPTER 2

The Social Meaning of Cartesianism: From the Self to Nature (and Back to the State)

I do not refuse to admit intellectual memory: it does exist. When, for example, on hearing that the word K-I-N-G signifies supreme power, I commit this to my memory, it must be the intellectual memory that makes this possible. For there is certainly no relationship between the four letters K-I-N-G and their meaning, which would enable me to derive the meaning from the letters.

> *Descartes' Conversation with Burman,* ed. J. Cottingham (Oxford, 1976), p. 9

Under the name of ethics I have included all the precepts that reason, that the state and that Christianity give for the regulating of human conduct either for themselves or for God: thus morality implies physics [or natural philosophy]; physics implies metaphysics; and metaphysics implies logic: and by this means all the parts of philosophy have such rapport and such links together as to be a whole that we can justly define . . . as the general system of philosophy.

> Pierre S. Regis, *Système de Philosophie* (Paris, 1690, *avec privilege du roy*), dedicatory preface to abbé de Louvois

SKEPTICISM AND NATURALISM

Galileo's confrontation with the church and its Aristotelian leadership was only one symptom of what had become a wider, more universal crisis over which philosophy commanded absolute authority in intellectual matters, as well as over who should establish the criteria by which that authority might be asserted. At stake was centuries of received wisdom, the synthesis of Aristotle and Christianity we describe as scholasticism, through which all church doctrines were explicated. The destruction of such a powerful and culturally unifying body of knowledge threatened to unleash socially dangerous beliefs and values, some of ancient, others of popular origin, which had been condemned for centuries by all orthodox theologians and philosophers. Any synthesis of the mechanical philosophy with those heresies, either with naturalism of popular and ancient origin or with materialism, would provoke the authorities in either Protestant or Catholic Europe to attempt its repression. Within this context the post-Galileo proponents of the new science, who in most cases shared the values and assumptions of those authorities, searched for a way to understand nature mechanically that would render that mental construction ideologically advantageous, yet another guarantor of social order and political stability.

By the early seventeenth century, however, any ideological construction intended to secure and unify the polity seemed destined to fail, yet another victim of the doctrinal struggle between the Reformation and the Counter Reformation. A century of religious controversy and open warfare between Catholics and Protestants had left many a civilized observer convinced that the only alternative to that brutal intolerance was skepticism, a refusal to believe anything doctrinal with absolute certainty. Such skepticism among the educated elite was profoundly dangerous to the maintenance of order and stability in society as a whole. Nonbelief, when rigorous, systematic, and searching, was considered a threat to all orthodoxy; no institution is safe if people simply stop believing in the assumptions that justify its existence.

Yet, somewhat ironically, skepticism as a mode of thought and argumentation was used in the sixteenth century by Protestants and Catholics alike—always in refutation of the other. It also gained currency as the result of the reintroduction of ancient Greek skeptical writers, in particular Sextus Empiricus, whose writings appeared in print in Latin (1562), in English (around 1590), and in a privately circulated French translation of 1630.[1] Giordano Bruno, for instance, makes mention of this skepticism, which he claims to have observed among academicians in the course of his various European travels.

By far the most sophisticated version of skepticism was put

forward by a late sixteenth-century French layman and adviser to the king, Michel de Montaigne. In the midst of the French wars of religion of the 1570s Montaigne lost his faith in human reason, in its ability to know anything with certainty. With devastating pessimism he labeled reason a "puny weapon," and in so doing his purpose was "to crush and trample underfoot human arrogance and pride."[2] He pointed to the controversy surrounding Copernicus and identified the conflicting opinions found in the new science as one more reason to proclaim the human predicament, the futility of searching for actual truth.

Montaigne gave voice to a profound intellectual crisis provoked largely by the Reformation but also by the new science. In response to this skepticism would come the first intellectual synthesis of modern thought to rest entirely and in the first instance upon the individual's ability to know nature mathematically and experimentally and, therefore, upon the belief in its possible mastery. Both assumptions are central to the philosophy of René Descartes, and his philosophy stands as one of the most powerful ideological adaptations of scientific knowledge to occur in preindustrial Europe. Descartes's discovery of the uses to which science could be put to ensure the foundation of orthodoxy, both political and religious, became a viable alternative, a way of rejecting both the fashionable skepticism of Montaigne's generation and scholasticism.

We can conjure up portraits of such late sixteenth- and early seventeenth-century skeptics as Descartes may have personally encountered: someone capable of converting from one religion to another at will; or one cynical about the claim to divine sanction, or indeed any sanction, for kingly authority; or a practitioner of no particular ethical creed associated with doctrinal Christianity who in consequence sought to live solely according to the moral dictates of nature. In short, skepticism threatened to unleash a practical as well as theoretical naturalism, a way of living and thinking that contemporaries called *libertinage*. Naturalism was commonplace in early seventeenth-century France, just as it had been in late sixteenth-century Italy. In Toulouse, Guilio Cesare Vanini (d. 1619) spoke of nature as God and was burned at the stake for so doing; so too was another pagan naturalist, Fontanier, burned to death in Paris in 1622. In that same decade alchemists were banned from the city and a prominent *libertin* was put on trial. The chaos of the sixteenth century had unleashed a rich and dangerous diversity in French intellectual life. It had created an unprecedented challenge to Catholic orthodoxy as buttressed by Aristotelianism. Heterodoxy also challenged the absolute authority of the monarchy and thereby threatened the sovereignty of the state.

Among the lesser French aristocracy and bourgeoisie, and even among the clergy, progressive elements were convinced that only strong monarchy could check the excesses of the nobility, *les grands*— many of them ultramontane or even pro-Spanish —as well as weaken the independent Protestant communities still powerful in certain districts and cities. For such monarchical Catholics as Father Mersenne, one of the earliest Parisian advocates of the new mechanical science, a new philosophical foundation for orthodoxy was now also regarded as absolutely essential. For Mersenne religion ensured the well-being of the state; indeed he believed, as he put it, that there should be *gendarmes temporels et spirituels* to enforce it. He and his mechanistically inclined friends, not least of them Descartes, repudiated the naturalists as well as the Aristotelians and sought in the new science the foundation on which might rest a new Christian orthodoxy. That quest, undertaken amid the fear of continuing political instability, gave a ferocious intensity to French natural philosophical debates of the early seventeenth century. These debates must also be seen as waged within terms set by the skeptics. They had praised the new science insofar as it challenged Aristotle and then went on somewhat perversely to use the achievements of Galileo to argue that nothing in science could lay claim to certainty.

These French skeptical attacks on the validity of science occurred precisely at the time when scientific learning aroused unprecedented interest among the educated laity. By 1632 a curious Parisian institution had developed to cater to that interest. This center, or Bureau d'addresse as it was called, was founded by one Theophraste Renaudot (d. 1653), a servant of Richelieu, a minor bureaucrat in the centralizing state that Richelieu struggled to create, a publisher, man-about-town, and devotee of the new science. At his Bureau minor noblemen jostled at weekly meetings with merchants, bankers, and tradesmen and shared their common enthusiasm for all new knowledge of a practical sort, and for science in particular.

To our great benefit Renaudot published accounts of those gatherings that reveal the extraordinary eclecticism—for its detractors the extraordinary confusion—that reigned in the minds of the cultured classes over ways and means by which nature might be understood.[3] What was being sought was some sort of "correct" method by which the natural world might be explored. Still dominant were the Aristotelians; but they were joined in debate by cabalists, believers in number mysticism, as well as by Paracelsians. These critics of the medical profession were at these Parisian meetings in great numbers, and they had taken up the ideas of the sixteenth-century German medical reformer Paracelsus. He had argued for natural remedies, for consultation with the stars, for a return to the healing power of nature in opposition to the blood-letting practices of the official doctors and the high fees commonly charged by them. At the Bureau some interest was even shown in applied mechanics, but it was the sciences useful for health or for trade that interested these bourgeois, as well as noble, seekers of truth. Before news of the condemnation of Galileo reached Paris, visitors to the Bureau even argued for Copernican heliocentricity.[4]

What these weekly meetings reveal is the existence of a market for science in early seventeenth century Paris, and they also remind us that natural philosophical language had now entered common parlance. Within that educated discourse, Aristotelianism, although still commonplace, aroused profound dissatisfaction; but no coherent alternative—except skepticism—existed to replace it. For those who would have orthodoxy in religion and order in the state that was a very dangerous state of affairs.

GASSENDI

Among the young men who came of age after the civil wars, but who had every reason to fear for the future stability of France after the assassination of its most able king, Henry IV (d. 1610), was a brilliant young priest from Provence, Pierre Gassendi (1592–1655). As a college lecturer there by the age of sixteen, Gassendi was one of the most gifted men of his generation. In a

letter of 1621 to one of the counselors of Henry's son, the young
and largely ineffectual King Louis XIII, Gassendi brooded over
"the great darkness of these times," over his fears, rooted "in the
history of France," that events were once again headed in the
direction of instability and possibly even civil war. Gassendi's
fears for the future gave him little confidence in the truisms of
the present, in either the philosophies of the people or those of
the "common run of philosophers."[5] And since "the Aristotelians far surpass all the others in number and obstinacy," they
merited, and received, Gassendi's particular and acute philosophical scorn. In his search for an entirely new foundation for
certainty, Gassendi assumed that the truth he would discover
must be beyond "the comprehension of common men . . . what
could a jackass do with a lyre?"[6] as he put it. He sought likewise
"to dig underneath the foundations" of Aristotelianism until its
"whole defensive system of walls and towers" collapsed.[7] His line
of attack took two forms: he used the now fashionable skepticism
drawn from the ancients to ridicule Aristotelianism, and he also
sought by the experience of nature to anchor science upon a new,
more empirical foundation. He sought to imitate practical men
—as he put it, politicians and farmers know what they see[8]—in
order to escape the logical syllogisms and obfuscations of the
scholastics. Gassendi, the priest, embraced the advocacy of reform in opposition to the scholastic clergy, and in so doing he
became the first of his generation, before Descartes, to offer the
new science as the foundation for a new, but empirically rooted,
certainty.

Gassendi attacked the syllogism, that basic tool of scholastic
logic, as useless because experience could always yield excep-
tions to its generalizations. There is remarkable irony in Gassendi's skepticism: science must be rooted in experience but
experience can always deceive. All those necessary and final
causes that the Aristotelians invoked to explain what and why
nature appears as it does are just unknowable, and in a brilliant
passage Gassendi makes the distinction—so basic to the mechanical philosophy of Galileo—between apparent qualities and real
ones, that which we see, such as color, and that which an object
really is.[9] All those qualities that the scholastics would situate as
inherent in nature Gassendi would strip away, and in their place
he offered the atoms of Epicurus. Had Gassendi been merely

another critic of the Aristotelians he would probably not have stood above the din of the Parisian crowd. But because he returned to an ancient heresy, that of Epicurean atomism, and Christianized it, and thereby made it the foundation for a new natural and human science, Gassendi acquired a vast following in his day. His influence stretched over Europe, not least through his English interpreters, to the colleges of Cambridge, and his atomism can be found discussed in the student notebooks of the young Newton.[10] Yet by the late 1640s his contemporary, Descartes, outstripped him in fame and importance because of his brilliance as a mathematician, but also in part because he addressed his audience in French in sentences that excel for their clarity in contrast to Gassendi's ponderous and rambling Latin tomes. Gassendi eventually became Descartes's rival, but in the end Cartesianism swept all before it.

In the 1620s and 1630s, however, Gassendi captured the attention of those who sought a new foundation for Christian orthodoxy that would also permit a mechanical approach to nature. He proclaimed his atomic philosophy cautiously, in select circles, in long Latin tomes. The dangers to his personal reputation, if not his safety, were real. Despite these, Gassendi championed a version of Galileo's principle of inertia; as Gassendi put it, "horizontal movement, whatever its cause, would be perpetual if some other force did not intervene to divert the moving body."[11] He embraced Copernicus, and he offered actual experiments to prove the validity of heliocentricity as well as the existence of the void or vacuum, a basic premise in Epicurean atomism. He took his atomic and experimental philosophy into the Parisian salons of the cultured elite, and he acquired an influential group of followers. He gave "courses" in his atomistic philosophy to provincial governors,[12] and he promoted it at a time when the anti-Aristotelians were being condemned by the church. Gassendi offered his philosophy as the only alternative to Aristotle, whom, he claimed, fails to offer us "a life without vexation." Gassendi also offered an escape from skepticism by advocating a belief in atoms as the core of reality.

But his Epicureanism was more than natural philosophy; he drew from it ethical and political implications. As one commentator has wisely phrased it, Gassendi "rehabilitated the ancient atomist Epicurius and interpreted his writings to mean that outside the sphere of faith and revelation man is governed by his senses, that is, by a desire to avoid pain and seek pleasure."[13] The realism of the new science when coupled with Epicureanism could justify a life spent in the pursuit of pleasure and the avoidance of pain, a life more secular than it was otherworldly—in short, an ethic based on the cultivation of the individual's passions and interests. Such cultivation defied centuries of church teaching that had been intended to curb and repress the passions.[14] It also possessed political implications of which Gassendi was well aware.

The pursuit of pleasure could lead, Gassendi feared, to political passivity, to a shirking of the responsibility shared by all humanistically educated members of the elite to participate in the life of the state, to give active allegiance to the source of order and stability, provided it is not tyrannical. To avoid the dangers presented by that sort of moral irresponsibility Gassendi equated pleasure with virtue. He defined reason as being simply the ability to perceive what will bring pleasure or avoid pain as well as being the good sense to mold our actions to that knowledge.[15] This active pursuit of one's passions and interests begins even before the institution of civil society; indeed it is the cause of that institution. Without government men would never be able to secure that which they covet and possess. The atomist brings to the state the same realism and empiricism that he brings to nature. He chooses to give allegiance to a king because having one supreme power makes the most sense. Provided the king eschews tyranny and gives offices to the aristocracy, Gassendi offers him a whole-hearted allegiance.[16]

From any traditional point of view, however, Gassendi's scientifically relevant Epicureanism could easily be perceived as dangerous to both church and state. Although it supported the role of divine providence in directing the motion of all material atoms, as well as the actions of kings, it did not endorse notions of the divine right of kings, nor did it offer the slightest consolation to clerical Aristotelianism. Gassendi sought a synthesis of the scientific investigation of nature with a new ethical posture that offered support to monarchy and strong central government provided they did nothing to inhibit the individual's free pursuit of his legitimate passions. In its intentions, although not in its details, this synthesis closely resembles that sought and found by

Gassendi's contemporary and critic, René Descartes. In the first half of the seventeenth century French natural philosophers sought a science grounded on philosophical assumptions that would permit the state to flourish to their mutual advantage and that would secure orthodoxy, although not necessarily the monopoly on truth claimed by the clergy entrenched in the schools and universities.

Not surprisingly, Christian advocates of the new science such as Gassendi had a difficult road to hold in seventeenth-century Paris. The accusation of heresy was made against him and his followers, and although he managed not to be prosecuted, the charge was never retracted. It is little wonder that Gassendi was cautious and resorted to ponderous Latin tomes wherein to develop his new philosophy, or that Descartes, as master stylist and persuader, chose to live much of his mature life in The Netherlands.

The systematic Aristotelianism of the scholastics was simply irreconcilable to the new science, either to the doctrines of Copernicus or to the mechanics of Galileo. That extraordinary intellectual achievement would have remained anathema to orthodox Christianity, and to that extent incapable of assimilation into the mainstream of European high culture, had the Parisian natural philosophers of the late 1620s and beyond not initiated a philosophical revolution that transformed mechanical and mathematical science into the foundation of an entirely new understanding of nature with direct implications for human institutions.

The question remains, why did that philosophical revolution occur in France and not in either Italy or The Netherlands, the only two other places where the new mechanical science had reached such a degree of experimental maturity? In other words, why was it Gassendi, and more importantly Descartes, and not the great mechanical scientist of the Low Countries, Isaac Beeckman (1588–1637)—who influenced them both profoundly—who were to achieve the philosophical synthesis that made Descartes, in particular, the greatest natural philosopher and advocate of the new science to be found in his generation? Before we can appreciate Descartes's achievement, which rendered mechanical science but one aspect of an entirely new foundation for all human inquiry as well as the source of a new cosmic order, we should examine briefly the thought of Beeckman, the only other mechanical philosopher outside of Counter Reformation Europe who might have been capable of producing such a grand natural philosophical synthesis based on mechanical principles.

BEECKMAN AND THE MECHANICAL PHILOSOPHY IN THE NETHERLANDS

Isaac Beeckman must be recognized as the first mechanical philosopher of the Scientific Revolution.[17] There were other mechanists before and contemporary with him—not least of them Galileo—but none of these developed a systematic philosophical approach to mechanical problems, one that speculated as to the atomic construction of matter and designated this mechanical philosophy of contact between bodies as the key to all natural forces, to every aspect of reality from watermills to musical sound. When Descartes first met Beeckman in the Dutch town of Breda in 1618 the French philosopher quickly came to recognize him as "his master."

Yet Beeckman never developed his mechanical philosophy into an entire philosophy for thinking, indeed for living, as Descartes was to do. We might speculate that this sort of grand cosmology was temperamentally alien to Beeckman and leave the matter at that. But to do so would be to dismiss the very real social, religious, and political differences—that is, the contextual factors—which separated the Dutch cities of Beeckman's time from the Paris of Gassendi, Mersenne, and Descartes. In the first instance Beeckman was a Protestant in a republic now dominated, after its rebellion against Spain, by a Calvinist clergy and, more important, by a Calvinist lay magistracy that exercised considerable authority over the clergy.

Beeckman's own religiosity was intensely pietistic and individualistic, close indeed to the spirituality found among the more extreme English Puritans* of the early seventeenth century.[18] His extreme Protestantism gave him absolute confidence that "God had so constructed the whole of nature that our understanding . . . may thoroughly penetrate all the things on this

*See glossary of terms.

earth.''[19] Beeckman never struggled with the fear of atheism when he approached either atomism or the mechanical philosophy; his Calvinism saved him from the struggle that gave birth to Gassendi's complex synthesis. Equally important, Beeckman encountered Aristotelianism in the Dutch schools and universities, but never did the Calvinist clergy entrenched in those universities possess the monopolistic power enjoyed by their counterparts at the Sorbonne. In a Dutch context one did not have to construct an entirely new foundation for learning in order to salvage Christian orthodoxy from the pretensions of the clergy, nor did Beeckman have to fear that intellectual dissent would literally destroy the Dutch polity or himself. Whatever profound disagreements separated Dutch Calvinists from one another in the early seventeenth century, neither clergy nor magistracy sought to shore up a monolithic state power as the only alternative to internal chaos. Indeed quite the reverse is true. The stability of the new republic rested very much on local and urban power in the hands of merchants and noblemen who had recently managed to free themselves from just such a monolithic and imperial power.

There is one other particularly relevant aspect of the Dutch situation that permitted Beeckman to develop his mechanical interests to the fullest. The Dutch cities were the most industrial in the known world; Beeckman himself came from a manufacturing family, and as a prosperous candlemaker turned mechanist, he easily mixed with merchants, navigators, and doctors. Among this laity he could meet in his own private mechanical "society" —*het collegium mechanicum*—where both practical and learned men could apply their mechanical interests to watermills or the problems of navigation, and this at a time when Dutch industry and commerce were everywhere expanding.[20] Those very practical men to whom Descartes was to make his appeal in France in the 1630s were already well established and secure in both an economic and political sense in the early Dutch republic.

This is not to say that contextual factors explain why Descartes wrote as he did, and Beeckman did not, but they did encourage, or demand, different responses. In that sense they permitted Beeckman to get on with his experiments, perhaps without ever forcing him to come to terms with the larger issues raised by the mechanical philosophy, the issues of stability and orthodoxy that confronted both Gassendi and Descartes.

CARTESIANISM

Certain unique conditions found in early seventeenth-century French society fostered the Cartesian synthesis. Those conditions permitted, indeed demanded, an intellectual revolution of the sort Descartes provided if they were to find resolution. The all-pervasive concern to restore certainty in learning without encouraging the monopoly enjoyed by the scholastic clergy, and the need to provide a new foundation for ethical and political conduct supportive of the central government, were those critically important factors. Without them, no long-term stability would be possible.

Both Descartes and Gassendi spoke primarily to secular elites, to offer science as an ally of their interests and passions and to bend both this new learning and these conflicting interests in the service of strong central government. The alternative to royal absolutism in early seventeenth-century France appeared to be the chaos of religious hatred and civil war. In the face of these possibilities Gassendi initiated the search for intellectual order; yet it was Descartes who achieved the synthesis that set the new mechanical science into a framework wherein it could be embraced not as heresy but as a profound truth, and as one incapable of creating social and intellectual disorder. Amid the maze of skeptics, *libertins,* naturalists, and hermeticists, Descartes cut a wedge that allied the new science with the individual's ability to will the attainment of his or her own knowledge. His proclamation, "I think, therefore I am," when allied to the support he and his followers self-consciously offered to central government, in short to absolute monarchy, allied science to the social issues of order and stability.

As a boy Descartes had received the finest scholastic and Jesuitical education available in the period, including, it would seem, in 1611 an account given at his school of Galileo's discoveries. He had also (and primarily) learned there that only ordered study, memory, and constant disputation according to the rules of logic produced true knowledge. Yet as a young adult he repudiated those approaches, concentrating his intellectual resources instead on mathematics, mechanical experimentation, and mathematical physics as pursued by Beeckman, and on devising a "new method" by which all knowledge might be reconstituted and freed from the assault of skepticism. The condemnation of Galileo in 1633 sent Descartes back to Holland in search of intellectual freedom and companionship. It was there in 1637 that he published his first major work, *The Discourse on Method,* which served as the preface to longer scientific treatises, one being the *Dioptrics,* the others being on meteorology and geometry.

Descartes believed himself to have a sacred mission to revise completely the accepted methods of learning and to establish the method of mathematical reasoning as the key to all learning.[21] In keeping with that mission—which came to him, he tells us, in a dream in the year 1619—Descartes, long before he published the *Discourse,* had come to define matter as would the mathematician. It is simply extension, and all qualities such as color and weight are simply accidents, as it were, that result from the size or relative motion of matter. The mind must grasp those abstract configurations of matter clearly and distinctly, just as the mathematician conceives of simple numbers and curves, or as the practitioner of the mechanical arts approaches a problem in simple, local motion. Such an approach to nature, mathematical in origin and deductive in application, was irreconcilable with the formal, often memorized education offered in the schools or with the scholastic definition of matter which held that qualities were inherent in bodies. The philosophical revolution that stands as central to the creation of modern science—namely the mechanical conception of nature—had already occurred in Descartes's mind before he sought to convert his audience in the *Discourse.*

The Discourse on Method

If we ask ourselves whom Descartes sought to address when he published the *Discourse* (in 1637 in Holland, then in Paris), we might find one answer in the scientific treatises to which that essay was intended as a preface. The *Dioptrics,* for example, aroused the hostility of later seventeenth-century scientists because they believed that Descartes had not in fact demonstrated in that treatise the laws of reflection and refraction which he there proclaimed. But those objections obscure Descartes's intention. Operating with the assumption that light is instantly transmitted, Descartes simply proclaimed the reflection and refraction of light, reasoning by mechanical analogies with bouncing balls and other moving bodies. But the treatise contained much more than those laws for which it eventually became famous. It dealt with light, the eye, the senses, the way the retina forms images, telescopes, and not least the method for cutting lenses. In short it was aimed initially at practicing, intelligent lens grinders.[22] As Descartes put it:

> The execution of the things that I shall say must depend on the work of artisans, who ordinarily have not studied at all. I shall attempt to make myself intelligible to everyone, and to omit nothing, nor to assume anything that one might have learned from the other "sciences."[23]

In this treatise Descartes wrote for an audience that had not been trained at the scholastic *collèges* or that, had it been, was also quite possibly disaffected from those methods of learning. In this essay Descartes sought to convert practical, but educated men of business and trade, among others, to the new mechanical philosophy, indeed to the new method of thinking about which they would have just read in the preface. While he himself moved largely in the circles of the minor nobility, among the polite and erudite of Paris, he sought also to capture the attention of just the sort of men who could have frequented the Bureau d'addresse. These were precisely the same groups to whom the policies of the absolutist kings of this era, Louis XIII and later Louis XIV, sought to appeal without at the same time dangerously alienating the old feudal elites.[24] It was an intelligent policy aimed at an audience that could appreciate the benefits to be derived from stability and expanding commercial activity.

The level of scientific awareness to be found among educated men of practical affairs in northern and western Europe from Descartes's time onward may have been remarkably sophisticated, although in most cases not very mathematical. Their interest in science that could be applied but that was also theoretical, rivaled, if not preceded, the scientific awareness achieved by the more leisured elites in the course of the seventeenth century.

As early as 1718 merchants in Amsterdam—more precisely a group of apothecaries, lens grinders, jewelers, and general merchants—were receiving lessons in Newtonian optics (only just published in 1704) as well as in Descartes's theory of light, Boyle's chemistry, and the new mechanics. They had specifically requested that such lectures be given.[25]

For such readers the Cartesian message is found most clearly articulated in the *Discourse* and its companion treatises. It proclaimed the self as the first arbitrator of knowledge, and Descartes presented that argument with all the rhetorical brilliance of a new prophet intent on proclaiming this new science. He begins the *Discourse* by proclaiming the relative equality of common sense, that most universal of human attributes. His appeal is directly to men of common sense; indeed he assures them that his own mind is really quite "ordinary."[26] Taking a swipe at the scholastics, he notes that even they had to admit this basic human equality in *forms* or *natures,* but not in *accidents.* Descartes as a mechanist will now proceed to eliminate even *accidents* from nature and philosophy.

Despite being of "ordinary" mind Descartes informs his readers that he has found a new method "to increase my knowledge," that he has made actual "progress . . . in the search for truth," but that he will not presume "to teach" it, only "to demonstrate how I have tried to conduct my own" reason (p. 28). Descartes eschews the pedantic role of the clerical schoolman,* so by "my frankness [I] will be well received by all." He then proceeds to demolish the learning of his youth, even though he was at "one of the most famous schools in Europe." All that he learned in letters and in "occult and rare sciences" did not in fact give him "clear and positive knowledge of everything useful in life." Of course, he performed some valuable intellectual exercises, but in the end he "was assailed by so many doubts and errors." Descartes appeals directly to the skeptics; indeed he even identifies with their plight.

The way out of this morass of pedantry and skepticism lies in latching onto that which will satisfy "the curious" (i.e., those who enjoy intellectual life for its own sake) and which will also "lessen man's work" (p. 30). Mathematics demonstrates "the certainty

*See glossary of terms.

and self-evidence of its reasonings," and not least it is useful for the mechanical arts. It can discipline the common sense we all possess; it can teach us to make our thoughts "clear and intelligible." It alone is perfectly suited to one who must think about matters "that affect him closely," and its reasonings are far superior to those taught by the men in their studies who "produce no concrete effect" and are out of touch with common sense (p. 33). Theology, by the by, can be left to those who have "some special grace from heaven" and not to "a mere man" such as Descartes. Likewise traditional philosophy is far too "subject to disagreement, and consequently uncertain" (p. 32). As for alchemy, astrology, and magic, with their naturalist associations, they are best left to "those who profess to know more than they do."

Descartes did not just stumble upon his method; first he had to explore "the great book of the world," the courts, the armies, where as a minor nobleman shielded by rank and education he met "people of different humors and ranks" (p. 33). Out of lived experience Descartes came to rely only on himself and his mind, disciplined by mathematics. In one of the most powerful metaphors in the *Discourse,* Descartes repudiates the wisdom of the ages, comparing it to those "old cities" built on the foundations of ancient and medieval ruins. With a vision one imagines as shaped by the ordered and relatively new cities of the Dutch Republic, with their geometrical and planned regularity, Descartes would have us build cities designed as those cities might have appeared to him, by "a single architect," "by the human will operating according to reason" (p. 35).

The Cartesian man whom Descartes would fashion, wills to use his reason, to create that which he wants for himself and "to lessen man's work." Sensing now that he is on dangerous ground in relation to the needs of public order, that such a man might not bend his will or reason to the state, Descartes admonishes his readers that only the laws of God and the state have made Europeans civilized. Consequently it is "unreasonable for an individual to conceive the plan of reforming a State by changing everything from the foundations"—how senseless it would be to demolish "all the houses in a town for the sole purpose of rebuilding them"—rather, the Cartesian method aims solely at ordering the individual's life (pp. 36–37). However difficult that personal quest may be, it is far easier than the difficulties that arise "in the reformation of the least things affecting the State" (p. 37). Only "meddling and restless spirits" in possession of "neither [the] birth nor [the] fortune to administer public affairs" (p. 38) are forever plotting the reform of the state, and Descartes makes it quite plain that none "could suspect me of this madness" (p. 38). Very few individuals are capable of the disciplined thinking that Descartes has willed for himself—the vast majority are either confused or simply follow "the opinions of others" (pp. 38–39).

It is as if Descartes is calling upon those restless men who might be tempted to reform the absolutist state to join him in an enterprise of quite a different sort, to rebuild solely the foundations of their own minds, "to try one's hand at architecture" (p. 45) of a most radical sort. To reorder the mind, to bend the will to conquer one's self, "to change my desires rather than the order of the world" (p. 47)—this is the task at hand. The reward Descartes promises for those who follow his scientific method is nothing less than mastery over nature. By comparison, he would have us believe, the alteration of the imperfections of the state is a paltry matter, and a dangerous one. Descartes saw, perhaps earlier than any other European outside Italy, that science in the right hands promised order and progress in the material realm without threatening to unleash the disorder that the early modern states dreaded above all else.

Descartes offers simple rules to guide these new bold spirits in their quest for a scientific mastery over nature: "Never accept anything as true that I did not know to be evidently so," avoid prejudice, include in your reasonings only that which presents "itself so clearly and so distinctly" that it cannot be doubted; that is, focus on real objects or on rules that explain their workings, order your priorities, begin with the simple and go to the complex, and proceed to impose such an apparent order even if none actually exists, and keep complete records and lists of what you are doing (p. 41). The method is both scientific and rational, although not rigorously experimental by post-Newtonian standards, and it stands as the first clear articulation of the new scientific methodology to be found in modern Western thought. This model for intellectual clarity depends heavily on Descartes's experience as a mathematician, who "keeps the right order for one thing to be deduced from that which precedes it" (p. 41).

Hence deduction, rather than induction based on experience or experiment, became the hallmark of seventeenth-century Cartesianism. That is not to say that Cartesians could not be experimental, as they were in late seventeenth-century Italy,[27] but they could also remain exclusively theoretical, as did many French Cartesians.

Perhaps the most stunning aspect of Descartes's way of scientific thinking was the radical charter it gave to the individual. While seeking always "to obey the laws and customs of my country," Descartes, and presumably those who would follow him, must doubt all other intellectual authority. Only the self, more precisely, the thinking mind—"I think, therefore I am"—can be taken as given. The first obligation of the scientific person is to embark on an intellectual odyssey that begins in doubt and ends with an affirmation of self. Interestingly, Descartes chooses to undertake that odyssey, as he tells his readers, in The Netherlands, where the society is already highly disciplined (p. 51), and where "busy people more concerned with their own business than curious about that of others" permit the philosopher to live in peace. The long shadow of Galileo's condemnation had touched Descartes in Paris, and he chose to move north; he also seems to have found the freedom enjoyed in the Dutch commercial cities much to his liking.

One of the most startling aspects of Cartesian individualism is its insistence that even the idea of God must first be perceived in the human mind before the being it describes can be acknowledged as real. Descartes's personal theism is manifest throughout his writings, but his method for affirming God's existence left little necessity for taking the preaching of ecclesiastical authorities as the main source of the individual's religiosity. Equally disturbing to those authorities was the tendency within Descartes's own thought and that of some of his followers to treat all of nature, including the operations of the human body, as explainable solely by reference to mechanical laws (p. 62). None of the "forms" or "qualities" found in scholastic definitions of matter are permitted; nature is solely matter in motion; from the circulation of the blood to the movement of light, "the rules of mechanics . . . are the same as the rules of nature" (p. 72). Only the existence of the soul renders human beings separate from that material order. Quite inadvertently, Descartes's radical separation of mind from body left the door open for a scientifically based materialism. Not least, Cartesian thinking about the deity left little room for the emotional experience of the divine so commonplace in early modern enthusiasm.

The Reception of Cartesianism

While Descartes's immediate followers, such as Malebranche and Regis, used his system in support of orthodox Christianity and rationally justified royal absolutism, Descartes's science, as Newton was quick to point out, opened the door to nothing less than a scientifically justified atheism. Far more dangerous than the naturalism that the Parisian circle of Mersenne or the theologians of the Roman Inquisition had sought to combat, this new naturalism, or scientific materialism, haunted the new mechanical philosophies of the later seventeenth century. The fear that Cartesianism ultimately led in a materialist direction became one of the main reasons for its ultimate rejection, particularly in England (see pp. 86–89), and for the concomitant acceptance of the Newtonian synthesis.

But for all its failings—not least its tendency to favor deduction over induction—Descartes's vision of the new science captured the attention of those groups in French society, and eventually elsewhere, who sought a radical alternative to scholasticism. In the words of one modern commentator, the Cartesian emphasis on the *cogito*—"I think"—made "a claim for human self-sufficiency of the most radical kind."[28] It was a claim that embroiled Cartesian science in decades of clerical opposition led by the scholastics. It only gradually emerged as triumphant, particularly in the schools and universities outside Catholic domination. Increasingly, educated elites accepted Descartes's proclamation that the new science grounded on mathematics and mechanical laws proved that

it is possible to arrive at knowledge which is most useful in life, and that, instead of speculative philosophy taught in the Schools, a practical philosophy can be found by which, knowing the power and the effects of fire, water, air, the stars, the heavens and all the other bodies which surround us, as distinctly as we know the various trades of our craftsmen, we might put them in the same way to all the uses for which they are appropriate, and thereby make ourselves, as it were, masters and possessors of nature. (*Discourse*, p. 78)

The reception of Cartesianism in France after Descartes's death in 1650 was delayed by the outbreak of civil war. Those very forces of decentralization, the clergy and *les grands,* coupled with the lesser nobility who resented the encroachments of absolutism, conspired for a time to plunge Paris, and elsewhere, into political chaos. The ensuing rebellion, known as the *Fronde,* dominated political and cultural life during the 1650s. Order was only gradually restored after 1661 and the assumption of personal control by Louis XIV.

The Truth Sought by Philosophers (1707). This engraving, a work of the early French Enlightenment, lionized Descartes, who, hand outstretched and led by Truth, in turn leads the ancient philosophers in their search. The figure being slain is Ignorance. Note that Aristotle has taken a back stage to Plato. The learned doctors of the Sorbonne thought this parade to be a thinly disguised attack on scholasticism. The engraver was forced to do a retraction in the form of another engraving entitled, "The Accord of Religion with Philosophy." In later eighteenth-century copies of the above engraving, Newton was substituted for Descartes.

As a result of that turmoil during the 1650s Descartes's science and philosophy made its greatest impact in the Dutch republic and in England (see pp. 86–87), among scientists as well as among the educated laity. The Dutch astronomer Christiaan Huygens produced a work of sophisticated astronomy, the *Saturnian System* (1659), which provided a quantitative and elegant description of the motions of the earth and Saturn around the sun, and the motions of the ring about Saturn. His final explanation rested on Cartesian matter theory, which required contact action to explain motion and hence postulated the existence of material vortices, pools of whirling matter, wherein the bodies of the heavens—more precisely Saturn's ring—move. He had observed the ring with his telescope, and in allying this new empirical and astronomical breakthrough with Cartesianism, Huygens gave added prestige to the new philosophical system. For Huygens, as well as for Descartes's more devoted and less critical followers, his writings provided a program for study, a call to concentrate scientific interest on the rules of motion and the nature of matter.

Indeed Huygens's role in the institutionalization and legitimation of the new mechanical science in Paris proved singularly important. After the defeat of the Frondeurs, the new monarchy under Louis XIV's personal guidance sought to consolidate its position by appealing to those lesser nobility and men of commerce eager to promote stability in the service of their own interests. Under the direction of his first minister, Colbert, science achieved a place in that grand design, and the founding of the Academy of Science was the first step in the process. Huygens explained to Colbert that the new mechanics with its emphasis on experiments concerning "the vacuum, force of powders [for gunnery], steam, wind and collision" could produce general laws, and that an academy was needed to promote this enterprise. The projects of such an academy would fulfill the dreams of Francis Bacon, and Colbert responded enthusiastically by establishing an academy charged with the task, among others, of developing the industrial arts and crafts.[29] The art of war figured prominently in the motives of the French monarchy, and the new mechanists never hesitated to assist in turning it into a science. Indeed the close linkage between the application of the new mechanics for civil and military engineering and the needs of the central government remained one of the dominant characteristics of French scientific practice until the French Revolution.

The cultural era associated with the reign of Louis XIV (1661–1715) combined grandeur with excellence in the pictorial and literary arts. New academies for the arts, architecture, and music were founded. All were intended to glorify the Sun King, and their autonomy was severely limited by royal ministers. Gradually royal patronage and the enthusiasm it had induced gave way to repression in those areas; this was especially so in publishing, where the freedom of the press threatened to undermine royal authority.[30] None of these pressures inhibited the Academy of Science in its pursuits or dampened the ardor of its Cartesian leadership for the policies of Colbert.

Cartesianism, as explicated by the first generation of Descartes's followers, gave ideological support to that linkage between science and state power. Not least, competition against the growing power of English science, also institutionalized in 1663 in the Royal Society of London,[31] further increased the importance accorded to science in the regime of the Sun King. Cartesian science, it was argued by the prominent Cartesian Jacques Rouhault, also further discredited the errors of "the vulgar,"[32] as well as of the Aristotelians, and it deserved a unique place within the culture of the educated elite. Knowledge of cosmology assists the study of geography and hence navigation and trade, while it is also essential that men understand the nature of metals, minerals, and salts, as well as medicine. Such sciences were directly suited to the needs of commercial capitalism and to the policies of Colbert. That Cartesianism might undermine Catholic doctrines, such as transubstantiation,[33] seemed irrelevant to the consensus about science promoted by Colbert and the Academy. In 1661 the right to censor all but religious books was removed from the doctors of theology at the Sorbonne and placed in the hands of the chancellor, who had the right to appoint royal censors.[34]

The most ideological and popular explication of Cartesian science was provided by a famous academician of the Academy of Science, Bernard de Fontenelle (d. 1757). His *Conversations on the Plurality of Worlds* (1686) went through five editions within four years of its publication, and some twenty-five editions in various languages by the mid-eighteenth century.[35] It presented a very simplified Cartesianism for the edification of the *noblesse* and *philosophes* of the salons, but it also specifically charged the Cartesian universe with the task of providing a model for the absolutist state. The scientific universe was to be the province now of the same elite that frequented the court and the salons, and knowledge of it permitted them to achieve a remote and superior knowledge of all other "worlds": those of "the people," and of all foreigners. Most important, Descartes's philosophy revealed an underlying order amid the apparent perturbations and dangerous motions that sometimes characterize the material order. Fontenelle would have similar laws work in society "which would fix people in those spheres of life that are natural to them." Just as the little planets obey the force of the larger planetary motions, so too, it is hoped, the smaller bodies in all places will render homage and respect to the larger preponderance of the state. It was an ideal that fitted well with the goals of the Academy as prescribed by Colbert. The scientist must be a disinterested servant of the state;[36] the *noblesse* should ponder closely his exemplary role.

It is worth noting at this point that Fontenelle's *Conversations* were addressed to an aristocratic lady. At the Parisian salons fascinated by the new science, we find the first evidence for a significant interest in that science among women. The Cartesian message of thinking systematically for oneself had the potential to enlist any educated human being as its follower. Although this interest in the new science among early modern women never provided them with access to institutional science and hence to power, their interest can be shown to have existed as early as the 1650s in both England and France and to have increased in the course of the eighteenth century.[37] Science, even when in the service of absolutism, could promote an intellectual freedom that ultimately boded ill for those who would monopolize its power.

With the consolidation of Cartesianism as a wholly new and comprehensive system for explaining the physical universe and for thinking scientifically, its earliest proponents sought through the printed word and public lecture to capture a new and even larger audience for science. This phenomenon of propagation, linked as it is to the printing press and the expanding world of secular fraternizing—the salons, coffeehouses, and cafés—was perhaps the single most important factor in rendering the new science into a unique and vital aspect of Western culture. It must be said that it was Cartesian propagandists and lecturers in the period prior to 1700, operating necessarily outside the traditional French schools and universities, who first sought the largest audience for science ever assembled in seventeenth-century Europe.[38]

A survey of that literature for the laity reveals much about the intentions of its authors and about the general tendency of Cartesian science to promote order in the state as well as commercial development. Following in the footsteps of Jacques Rouhault, Pierre Sylvain Regis lectured in Paris and the provinces and spread Descartes's message, along with the need for absolute authority and order in the state. Reason would guide the self-interested subject in this ordered probity, just as the laws of Cartesian physics guide the universe. For Regis, as the second quotation used to open this chapter indicates, physics and morality were intimately connected, and both must serve the cause of reason and order in society and government. His lectures moved effortlessly from an explication of the Copernican universe based on Cartesian principles, to the earth, the nature of air, water and salt, the properties of metals, fermentation, vegetation, simple chemistry in general, wind, the motion of the sea, the role of God in creation, anatomy, the principles of civil society, the nature of the passions, and only last, mechanical devices intended largely to illustrate Cartesian mechanics. In many respects these lectures, first published in 1690, anticipate the format adopted within a decade by the English and Dutch followers of Isaac Newton. But in one critical respect these Cartesian lectures differ from their Newtonian successors: relatively little attention was paid by Cartesian lecturers to mechanical devices intended for industrial application.

The vision and practical applications offered by the Cartesian lecturers of the late seventeenth century were essentially commercial, and their science addressed the needs of commercial capitalism. Although sometimes immensely sophisticated in hydrostatics or meteorology, they simply did not place emphasis on local motion and mechanical devices needed to teach the laity how to harness mechanical principles in the service of industry. The Cartesian natural philosophers of seventeenth-century Europe possessed in many cases a vision of the role of science far in advance even of that imagined by Colbert. But when they sought an audience, both philosophers and audience responded to the material order everywhere around them. To that extent Cartesian science would remain, in its mainstream, ideologically absolutist in politics, and practically commercial in application. In contrast, the science of the Newtonians was to become the science of constitutional monarchy and of early industrialization.

Yet the Cartesians were the true pioneers of modern science. They often labored in a hostile climate. The Sorbonne had little use for them; eventually the archbishop of Paris closed down Regis's lectures "in deference to the old philosophy." Despite this opposition Cartesian science spread in France, although not as rapidly as it did in Protestant Europe. Its French growth, however slow, eventually established Descartes's science at the Academy and in about 170 liberal arts colleges (out of 400) that actually taught natural philosophy. Once ensconced, Cartesianism proved intractable. By the 1740s, when Newtonian mechanics had captured the educational system in England, in Scotland, and at the major Dutch universities, the French colleges continued to teach Descartes relentlessly. As a result educated Frenchmen of the generation prior to the 1750s missed any formal education in practical Newtonian mechanics as well as in the entire Newtonian philosophical outlook (see Chapter 3).[39]

THE SPREAD OF CARTESIANISM

The history of the spread of Cartesianism is another palpable illustration of the ability of Protestant culture to absorb the new science more readily than the culture dominated by the Roman clergy. This is not to say that Protestants always trusted Descartes. Those of Aristotelian persuasion, such as the Utrecht professor Voet, as early as 1639 attacked Descartes for atheism. He saw the Cartesian will, that radical assertion of the individual's right to think, as a defiance of God's will and as a license for rapacious greed.[40] He also feared that such radical freedom could lead to "enthusiasm," which meant to any seventeenth-century supporter of order and authority the license claimed by radical sectaries to follow the light of their own conscience and, if need be, to spurn clerical or magisterial authority. In the seventeenth century men policed the mind before all else; the resources were simply not available to do more.

In mid-century in the Dutch city of Utrecht, the Calvinist opponents of Descartes sought to rally the people against his philosophy. His supporters identified the triumph of his science at the university as a clear sign of the power of the lay elite, the *regenten,* to control not only the church but the city in general. They sought to continue their profitable control over the ecclesiastical properties seized at the Reformation, as well as to permit the relative religious toleration on which Dutch trade and commerce flourished. At first the Cartesians had to proceed slowly. The lay magistrates feared public controversy that stirred up the populace even more than they disliked Voet. Yet by the late 1600s every Dutch university sported a Cartesian or two on its faculty. Despite the real fear of materialism voiced in some quarters, by the 1670s Cartesian science dominated the University of Leiden, perhaps the foremost scientific center on the Continent in this later period. When a physics laboratory was set up in Leiden it was used solely for the purpose of illustrating Cartesian principles, not for testing them.[41]

The fear of the materialistic implications of Cartesianism failed in the final analysis to halt its Continental progress, even in select Italian circles. For orthodox Calvinists of the Genevan variety, Cartesianism could even be used—despite Voet—to bolster fideism, to permit orthodoxy based on philosophy and not on theological disputation. The Genevan Cartesian Louis Tronchin used the Cartesian definition of matter as extension to support the Calvinist assault on transubstantiation, to argue that the spiritual Christ could not be physically present in the matter of the Eucharist.[42] In these late seventeenth-century Calvinist circles that adopted Cartesian science wholeheartedly, we can see a growing separation between theological discourse and philosophical convictions, not as yet a loss of faith, but simply a separation of the languages of explanation and discourse. Such separation could, however, lead to crises of confidence and hence of faith. By 1700 Cartesianism had contributed to what the French historian Paul Hazard labeled "the crisis of the European mind" (see Chapter 4), the decisive shift from a religious understanding of nature and human institutions to a secular one.

But the possibility of such a crisis was remote in the minds of most of Descartes's followers prior to 1700. In the 1670s at the University of Edinburgh, for example, Cartesianism was well established, although in the 1650s during the English Revolution the fear of atheism had threatened to inhibit that process.[43] Only in select philosophical circles in England, more precisely at the University of Cambridge in the 1660s, did the fear that Descartes's philosophy would lead directly to atheism (as the young Isaac Newton bluntly put it in his private papers) force Newton and his circle of mentors to break with the Cartesian version of the mechanical philosophy. They sought at all costs, however, to save the new and fragile mechanical tradition against the ever-present possibility of an Aristotelian revival. As a result, an intellectual revolution occurred in mid-seventeenth-century England that made the new science a central part of a new religious consensus. In English science from the mid-1650s onward, a degree of cultural integration was achieved that permits us to speak of the Anglican origins of modern science (see Chapter 3).

But the quest for such an integration had begun in Paris during the 1620s, and without Descartes and his Cartesian followers would have been very much delayed. Cartesianism sanctioned a mechanical approach to nature that made it the only plausible explanation open to those capable of thinking "clearly and distinctly." It offered the prospect of extraordinary material rewards, as well as the gratification of knowing that no authority dictated to the thinking self. The individual's disciplined reason alone sanctioned the very existence of those external authorities. In so doing, Cartesianism created an intellectual order in the service of self-interest and human desire that rigorously defended the ideological foundations of monarchical authority within an increasingly centralized state.

NOTES

1. For a splendid discussion of the roots of seventeenth-century skepticism, see Richard Popkin, *The History of Scepticism from Erasmus to Descartes* (New York: Harper & Row, 1964), Chapters 1–3.
2. Quoted in ibid., p. 46.
3. T. de Renaudot, ed., *Recueil general des questions traitées et conferences de Bureau d'addresse,* 5 vols. (Paris, 1658–1666).

4. See the excellent discussion in Geoffrey Vincent Sutton, "A Science for a Polite Society: Cartesian Natural Philosophy in Paris During the Reigns of Louis XIII and Louis XIV" (Ph.D. diss., Princeton University, 1982), Chapter 1. The opening pages of this chapter are indebted to this account.

5. Craig B. Brush, ed. and trans., *The Selected Works of Pierre Gassendi* (New York: Johnson reprint, 1972), pp. 4–5.

6. Ibid., p. 23.

7. Ibid., p. 26.

8. Ibid., pp. 34–35.

9. Popkin, *Scepticism*, p. 105.

10. J. E. McGuire and Martin Tamny, eds., *Certain Philosophical Questions: Newton's Trinity Notebook* (Cambridge: Cambridge University Press, 1983), pp. 20–21.

11. Brush, *Gassendi*, pp. 130–131.

12. See Pierre Gassendi, *Opera Omnia* (Lyon, 1658), vol. 6, pp. 116–160; and see Howard Jones, ed., *Pierre Gassendi's Institutio Logica (1658)* (Assen: van Gorcum, 1981). Cf. R. Lenoble, *Mersenne ou la naissance du mecanisme* (Paris: Vrin, 1943), pp. 157–167, 180, for context.

13. Lionel Rothkrug, *Opposition to Louis XIV: The Political and Social Origins of the French Enlightenment* (Princeton: Princeton University Press, 1965), pp. 83–84.

14. For a good example of that sort of traditional morality see R. P. I. F. Senault, *De l'usage de Passions* (Amsterdam, 1668).

15. Lisa Sarasohn, "Ethical and Political Philosophy of Gassendi," *Journal of the History of Philosophy*, vol. 20 (1982), pp. 239–240.

16. Ibid., pp. 252–254. For a discussion of the kind of instability that could be generated by the opposition clergy, see Richard M. Golden, *The Godly Rebellion: Parisian Cures and the Religious Fronde, 1652–1662* (Chapel Hill: University of North Carolina Press, 1981).

17. See Klaas van Berkel, *Isaac Beeckman (1588–1637) en de Mechanisering van het Wereldbeeld* (Amsterdam: Rodopi, 1983), p. 215. I am indebted to this work for my summary of Beeckman's career.

18. Ibid., pp. 21, 73–75.

19. Ibid., p. 164.

20. Ibid., p. 70.

21. A. J. Krailsheimer, *Studies in Self-Interest: Descartes to La Bruyère* (Oxford: Clarendon Press, 1962), p. 32.

22. I am very indebted in this discussion to Bruce Stansfield Eastwood, "Descartes on Refraction: Scientific Versus Rhetorical Method," *Isis*, vol. 75 (1984), pp. 481–502.

23. As quoted in ibid., p. 486.

24. A. D. Lublinskaya, *French Absolutism: The Crucial Phase, 1620–29* (Cambridge: Cambridge University Press, 1968), p. 33. Cf. Orest Ranum, *Artisans of Glory: Writers and Historical Thought in Seventeenth Century France* (Chapel Hill: University of North Carolina Press, 1980), p. 119, on Descartes as a scathing critic of *ars historica,* that

genre of historical writing which gloried in the heroic rather than in the rational vindication of royal authority. For an excellent discussion of recent scholarship on absolutism, see William Beik, *Absolutism and Society in Seventeenth-Century France* (Cambridge: Cambridge University Press, 1985), Chapter 1.

25. See the notes taken on lectures of Daniel G. Fahrenheit, 1718, University Library, Leiden, MS BPL.772. ff. 1–100. Cf. Pieter van der Star, ed. and trans., *Fahrenheit's Letters to Leibniz and Boerhaave* (Amsterdam: Rodopi, 1983), p. 105.

26. René Descartes, *Discourse on Method and the Meditations,* trans. F. E. Sutcliffe (Harmondsworth, U.K.: Penguin, 1979), p. 27. For the convenience of readers I am using this edition for my explication. All page numbers in the text henceforth refer to it.

27. Paula Zambelli, *La formazione filosofica di Antonio Genovesi* (Naples: Morano, 1972).

28. Krailsheimer, *Studies in Self-Interest,* p. 35. For a highly philosophical approach to the political implications of Cartesianism, see Pierre Guenancia, *Descartes et l'ordre politique* (Paris: Presses Universitaires de France, 1983). For a discussion of elite culture, see Robert Muchembled, *Popular Culture and Elite Culture in France 1400–1750,* trans. Lydia Cochrane (Baton Rouge: Louisiana State University Press, 1985), which is strangely silent about science.

29. James E. King, *Science and Rationalism in the Government of Louis XIV, 1661–83* (Baltimore: Johns Hopkins University Press, 1949), pp. 292–294.

30. See Joseph Klaits, *Printed Propaganda under Louis XIV: Absolute Monarchy and Public Opinion* (Princeton: Princeton University Press, 1976); cf. for a survey of the arts under Louis XIV, Antoine Adam, *Grandeur and Illusion: French Literature and Society 1600–1715* (London: Weidenfeld, 1972).

31. See Samuel Sorbiere, *Relation d'un voyage en Angleterre, où sont touchées plusiers choses, qui regardent l'état des sciences . . .* (Cologne, 1667), pp. 58, 69.

32. Jacques Rohault, *Traité de Physique* 2 vols. (Amsterdam, 1672), vol. 1, pp. 13–17; vol. 2, pp. 142–143.

33. [J. G. Pardies and Rochon], *Lettre d'un philosophe à un Cartesien de ses amis* (Paris, 1672), p. 5.

34. Erica Harth, *Ideology and Culture in Seventeenth-Century France* (Ithaca, N.Y.: Cornell University Press, 1983), pp. 162–163.

35. I owe this perceptive reading of the *Conversations* to Sutton, "A Science for a Polite Society," pp. 437–452.

36. Harth, *Ideology and Culture,* p. 231. Cf. M. de Cordemoy, *Dissertations physique sur le discernement du Corps et de l'Ame . . . sur le système de M. Descartes* (Paris, 3rd ed., 1689), vol. 2, *epitre.*

37. See Ruth Perry, "Radical Doubt and the Liberation of Women," *Eighteenth Century Studies,* vol. 18 (1985), pp. 472–493.

38. J. L. Heilbron, *Electricity in the 17th and 18th Centuries: A Study in Early*

Modern Physics (Berkeley: University of California Press, 1979), pp. 146–159.

39. L. W. B. Brookliss, "Newton, Descartes and the Mechanical Philosophy: Physics Teaching in the French *Colleges de Plein Exercise,* 1700–1789," paper presented at a meeting of the British History of Science Society, University of Essex, 1982.

40. Thomas Arthur McGahagan, "Cartesianism in The Netherlands, 1639–76: The New Science and the Calvinist Counter Reformation" (Ph.D. diss., University of Pennsylvania, 1976), pp. 154–160.

41. Ibid., pp. 335, 372. Cf. Th. H. L. Scheurleer and G. H. P. Meyjes, eds., *Leiden University in the Seventeenth Century* (Leiden: Brill, 1975), p. 312.

42. Michael Heyd, *Between Orthodoxy and the Enlightenment: Jean-Robert Chouet and the Introduction of Cartesian Science in the Academy of Geneva* (Nijhoff: The Hague, 1982), pp. 72–74.

43. Eric Forbes, "Philosophy and Science Teaching in the Seventeenth Century," in Gordon Donaldson, ed., *Four Centuries: Edinburgh University Life, 1583–1983* (Edinburgh: University of Edinburgh Press, 1983), pp. 31–35. Cf. Alan Gabbey, "Philosophia Cartesiana Triumphata: Henry More (1646–71)," in Thomas M. Lennon et al., eds., *Problems of Cartesianism* (Kingston, Ontario: McGill-Queen's University Press, 1982), pp. 244–250.

CHAPTER 3

Science in the Crucible of the English Revolution

By 1640 in England, Descartes's *Discourse* had been read and appreciated in select circles while the earlier writings of Francis Bacon (see pp. 32–36) were also undergoing a marked revival. Here, as in France, the issue remained that any dramatic alteration in the concept of the natural world required the integration of socially relevant beliefs and the needs of authority—the integration, in short, of religion and the political order. But in mid-seventeenth-century England the intellectual fermentation provoked by the new science occurred within the context of a larger political and religious agitation for reform and renewal.

The English reformers of the 1630s hurled their demands for reform against what they saw as an obstinate monarchy, a corrupt court, and an insufficiently Protestant church. This confrontation ended in political revolution, and that social upheaval in turn directly affected the direction taken by the new science. On the basis of modern scholarship, we can now say that no single event in the history of early modern Europe more profoundly shaped the integration of the new science into Western culture than did the English Revolution (1640–1660). It shaped the natural philosophical thinking of Robert Boyle (1627–1691) and Isaac Newton (1642–1727) in ways that assisted in the development of their purely experimental and mathematical interests; and out of their achievement, that synthesis of philosophy and experimental method we describe as modern science came to exist. In addition the revolution, with the Puritan reformers in its vanguard, raised the fundamental issue of the social uses of the new science; indeed scientific progress was central to their revolutionary vision. By the late 1650s they had failed to achieve their goals, but in the process they had rendered science and natural philosophy vital elements in any alternative social ideology. By 1660 and the end of the first phase of the English Revolution, the prosperity of the English state came to be seen as linked—at first tentatively and then decisively—with the development of science and technology.

Various historians writing over several decades have demonstrated that the English Revolution, understood as a crisis stretching from the 1640s to the late seventeenth century, bears close relation to the development of English science in that period.[1] In the 1930s the American sociologist Robert Merton illuminated the links between the progenitors of that revolution, namely the English Puritans, and the origins of modern science.[2] He provided convincing evidence to show that the Puritans, indebted as they were to the doctrines of striving and predestination, that is to Continental Calvinism, were also particularly attracted to scientific inquiry. This evidence further strengthened the case for the linkage between European Protestantism and the rise of science, a case that can be documented not only in England but also in the Dutch Republic after its liberation from Spain (1585).[3] More recent British and American historians— such as J. R. Jacob, Charles Webster, and Christopher Hill—have further illuminated the linkage between science and the Puritan reformers and stressed the importance accorded to science in the 1640s by a circle of those leading reformers led by Samuel Hartlib.

BACON AND THE PURITANS

These Puritan reformers seized on the writings of Francis Bacon as their guide to the new scientific spirit and its empirical methodology. They interpreted Bacon not as the humanist he was, but rather they emphasized the millenarian and reformist aspects of his thought as revealed in *The New Atlantis* as well as *The Advancement of Learning.* Perhaps the most unique aspect of reforming Protestant thought, and one that sets it off from many Continental religious movements that embraced the new science, is its millenarianism. Put quite simply, English Puritans believed literally in the scriptural prophecies about the final days and the end of the world. God directs the course of human affairs, just as he directs the course of nature. At some point in time, one that could be determined by careful scholarship or even possibly by a gnostic illumination, history and nature would synchronize as God destroyed the world in a cataclysmic upheaval that would precede the institution of a thousand-year reign of the saints, the millennium. Almost every important seventeenth-century English scientist or promoter of science from Robert Boyle to Isaac Newton believed in the approaching millennium, however cautious they may have been in assigning a date to its advent.

That extraordinary fact ought to alert us to beware of the simplistic categories—scientific versus magical, rational versus irrational—as having much relevance to the ideology of seventeenth-century science. Contrary to the retreat from worldly concerns or the conservatism that we might expect fundamentalist millenarianism to inspire, the vision of the Puritans made urgent the work of reform and renewal, in which science figured as a critical element. Hartlib and his friends promoted Baconian schemes for classification and improvement that led them to advocate universal education and medicine for everyone (a physician in every parish was the goal). They embraced the new philosophy as the cornerstone of all scientific inquiry; they championed mechanical experiments intended for industry, and chemical experimentation for agricultural improvement.

With the outbreak of civil war against the king, the 1640s became an exhilarating time for social reformers as well as for natural philosophers and scientific experimenters sympathetic to the parliamentary cause. From that period we can trace the inspiration for the eventual founding of the Royal Society of London (1662), as well as plans for the founding of new colleges, grammar schools, and academies, for a multitude of technological innovations in everything from mining to banking, and for a central "office of address" for the communication of useful knowledge.[4] We can locate the 1640s, too, as the time of the earliest chemical experiments of the young Robert Boyle, an associate of Hartlib and his circle.

Within this context established by revolution and civil war, modern science at its very origins was perceived in terms of its social usefulness and linked to a larger vision of reform and enlightenment. The Puritan reformers assaulted the old monopolies of the physicians and the universities, and they took up the latest in scientific inquiry—everything from the theories of Paracelsus to the writings of Galileo, Bacon, and Descartes.[5] Had they succeeded in the 1640s in establishing even a modicum of these reforms, particularly in the areas of medicine and scientific education, the more humane aspects of scientific inquiry, as distinct from military and commercial adaptations, might have come to prevail. But history cannot be written about what might have been, and by the 1650s a very different mood had come to prevail even among many of the reformers.

THE IMPACT OF THE REVOLUTION

Perhaps the greatest fear of any early modern elite focused on the danger of popular unrest. The widening gap in seventeenth-century Europe between rich and poor, or simply between the relatively prosperous and the indigent, coupled with the absence of effective policing mechanisms, made even the possibility of lower-class revolt, whether by peasant or artisan, the most dreaded of all possible occurrences. In France one of the mainstays of monarchical power became the expected ability of the king and his bureaucracy to ensure an adequate supply of bread in the cities and the countryside. Failure to do so threatened the possibility, or the reality, of civil unrest among the lower orders.

But in England in the 1640s policing mechanisms largely dissolved. The censorship of books, a function of established church censors, largely disappeared, as Parliament dismantled the privileges of the Anglican church* and failed to put in place an equally effective alternative. As a result more books and pamphlets were published in England from 1640 to 1660 than in the entire rest of the century. More dangerous than books, however, were the various new sectarian movements that came to the surface abruptly in the 1640s. The theology of many of these groups can be traced back to the more radical phase of the sixteenth-century Protestant Reformation, when doctrines such as the priesthood of all believers, or the enthusiastic notion of the "inner light" within every man and woman, came to justify the religiosity and religious freedom of lower-class, often illiterate, reformers. That sort of radicalism was not what the

*See glossary of terms.

Puritan reformers had in mind when they urged their reforms upon Parliament.

But Parliament came to be dependent in the war against the king on the New Model Army, and within its ranks could now be found many of these radical sectarian movements and their most articulate leaders. Suddenly the dialectics of revolution threw to the surface a threat more serious even than the absolutist policies of Charles I against which Parliament had initially revolted. The radical reformers and sects—the Levellers, Diggers, Ranters, Quakers, Muggletonians, to list only the most prominent— demanded a range of reforms, from one man, one vote, property redistribution, and complete religious toleration, to the right of women to preach, the end of church tithing, and the curtailment of the privileges of physicians and lawyers. Some radicals mocked the sober life style of the Puritans, their dedication to the work ethic, and their haughty strictures against swearing, smoking, drinking, and sexual freedom.

The threat posed to the moderate reformers included a direct challenge to the kind of science they would promote. Radical reformers also advocated a new science; they too repudiated the Aristotelian learning entrenched at Oxford and Cambridge. But in place of Aristotle they would put Paracelsus and the naturalism and magic associated with the Hermetic tradition. In the words of John Webster, a surgeon and for a time chaplain in the New Model Army, the radicals wanted "the philosophy of Hermes, revived by the Paracelsian Schools"; he would have installed in the universities "true Natural Magicians, that walk not in the external circumstance, but in the center of nature's hidden secrets." They must have "laboratories as well as libraries" in order to pursue their alchemical and medical experiments.[6] The promoters of the new mechanical philosophy took to the offense against the natural magicians and simple "mechanicks" (or artisans), or more precisely against the radical and sectarian forces that would "turn the world upside down" by using the "science" of the people presumably in their interest.[7] The moderate scientific reformers—such as Robert Boyle, William Petty, and John Wilkins, as well as Henry More, the Cambridge Platonist—led the assault on the radicals and used the new mechanical philosophy, particularly of Descartes and also Gassendi, against the naturalism of the magicians. In short these men, who had begun the revolution as scientific reformers, found themselves outflanked by the radicals and forced in consequence to articulate an intellectual and social stance that now sought to conserve as much as it sought to change.

THE REACTION AGAINST SECTARIAN RADICALISM

One group of moderates—led by Boyle, Wilkins, John Wallis, John Evelyn, Christopher Wren, and others—went on to become the founders of the new Royal Society.[8] Their goal was now to promote the organized pursuit of experimental science but to detach it from any attempt at the radical reform of church, state, the economy, or society. They did not cease entirely to be reformers but couched their reforming sentiments in vague terms of improving man's health and estate through science. When they did become more specific it was, for example, to indicate the ways by which experimental science might be deployed to increase production, especially food production, and commerce.[9] But these reforms would progress without altering the existing social arrangements in the direction of greater leveling or redistribution of property. This is not to discount these reforming aims but to indicate that they would not democratize politics and society.

As if to symbolize this approach to reform, many in this once moderate but now increasingly conservative group left revolutionary London and retreated into Oxford colleges away from the social and political turmoil, there to pursue their thoughts in quiet contemplation. And when radicalism threatened to jeopardize their freedom within the university, they stepped in to oppose the radicals.[10] These moderate reformers made a point of avoiding all religious and political questions while discussing science. This, however, does not mean that they were unaffected by the outside world and did not have their own opinions about it, but rather that their science was not intended to democratize power relations within society.[11] In this regard the reformers consciously distanced themselves from the radicals, who saw in science a powerful tool for promoting religious, political, and social revolution. In the radical vision science could justify democracy in church and state; it might also be used to extend popular education in schools and universities and to build a new society that would be more just and rational.

The moderate reformers did more than simply retreat to Oxford and defend it against the proposals of the radicals. What was at stake by the early 1650s after the defeat and execution of the king in 1649 was, as the moderates saw it, nothing less than the survival of the polity. The country was now governed by Parliament and the army, and the threat of lower-class and sectarian radicalism seemed everywhere apparent.

What must be grasped in this crisis is the important role of philosophies of nature in giving expression to human goals and aspirations. Natural philosophical and religious language informed discussions about the nature of political authority, the rights of the church, and the relationship between master and servant, husband and wife, lord and commoner. To picture the cosmic order, "the world natural" as Newton put it, was to speak, by analogy, of "the world politick." For a body to rise toward heaven or fall toward the earth might symbolize the "rising and falling in honour and power"[12] of states or individuals. On a more abstract level the relationship between God and nature, hence between spirit and matter, could express the beliefs of an individual or sect concerning the role of priestly or kingly authority. If God is remote from his creation, if spirit clearly dominates matter, does this not justify the continuation of that same authoritarian structure in society and government, both ecclesiastical and civil? Or to put it another way, if the spirit of God dwells in nature, in everyone, what need is there for the heavy hand of priestly or magisterial authority?

Such questions were directly relevant to the new science because its most pressing concern centered precisely on how to define the relationship between spiritual forces and matter. The Cartesian legacy demanded that matter be seen solely as extension, as the physical protrusion of an infinitude of corpuscles in space. Consequently its motion could be explained only by reference to contact with other matter. Such an explanation could easily counter arguments that depended on inherent spiritual qualities or occult forces, which in turn justify a magical approach to nature. In one sense the new mechanical science was a perfect tool to hurl against the hermetic and the magical. But in another sense, those very arguments came dangerously close to situating within matter the power to move itself. And once so endowed, how did that matter differ from the universe described by certain radical philosophers and sectaries as so filled with the spirit of God as to be moved by a pantheistic force open to the experience of every man and woman?

In England by the 1650s there were many such pantheistic philosophers openly publishing and preaching their beliefs. The Digger, and first English communist, Gerrard Winstanley, believed that God was in all things, that the creation was the clothing of God.[13] He also equated God with Reason—an idea drawn directly from the Hermetic tradition with which the sixteenth-century reformer Giordano Bruno had also been in contact (see pp. 26–28). Among other radical sectaries, the Ranters believed that hermetic and pantheistic doctrine, and with it they justified their self-conscious departure from Puritan morality. One of the leading Quakers of the period, George Fox, admitted that he had nearly succumbed to the doctrine, also preached by Ranters, that there is no God and all creation comes from nature.[14] The Leveller Richard Overton argued for a pantheistic materialism and for the doctrine that the soul falls asleep at death; other radicals believed that it died with the body. All of these ideas were part of a metaphysic intended as the foundation for a new religiosity and a new society, one less inequitable, freer, less rigid in its social and economic divisions, less priestly and magisterial in its system of authority. These ideas constituted a direct challenge to the authority of the propertied and educated elites.

HOBBES

Heresy during the English Revolution came, however, dressed in many guises and not least from the scientific community itself. Thomas Hobbes may best be known today as a political theorist; but in the 1640s Hobbes was at the center of scientific and natural philosophical discussions in Paris as well as London. He was a committed mechanist and a violent opponent of Aristotle and his followers. He was also a royalist* and in no sense a sympathizer with radical causes. Yet his political philosophy as found in his

*See glossary of terms.

most famous work, *Leviathan* (1651), rested on a mechanical philosophy that ignored entirely the operations of spiritual forces in nature. In denying the existence of spirit he likewise denied any independent role to the clergy—"those bugbears who prick the sides of their princes," as he described them—and they were the traditional guardians and interpreters of the operations of spirit in the world. The clergy, according to Hobbes, were fraudulent as intermediaries between God and man and should be reduced to mere functionaries of the civil sovereign. From the perspective of the Puritan moderates and Christian scientists, or *virtuosi,* * Hobbes and the radicals were curiously allied in their heretical philosophizing. The first sucked spirit out of the world so as to rationalize and explain solely in material terms the existence of human greed and self-interest; the second pumped it back in with the intention of shaking the material and social order.

During the 1650s the scientific reformers modified their understanding of nature to answer the threat posed by both Hobbes and the radicals. In the place of occultism and materialism, Robert Boyle offered what he called his corpuscular philosophy* and rendered it the foundation of the science of chemistry. This amounted to a Christianized Epicurean atomism,* borrowed in part from Gassendi, which Boyle afforded the status of a hypothesis to be tested by experiment, not that of a dogma. Boyle and his associates held with Epicurus that the world was made up of lifeless atoms colliding in the vacuum of space. But the Puritan philosophers and their associates, like Boyle, departed from Epicurus by denying that the world as we know it was the product of a long succession of random atomic collisions. Rather, they held that a providential God and not random chance was responsible for all motion in the universe. He determined the paths the atoms took and hence maintained the order of the universe.[15] Not only was this a workable scientific hypothesis capable of being refined and elaborated by a Baconian program of experiment, it was also an attractive candidate for adoption because it was applicable to social issues.

This Christianized corpuscular and experimental philosophy allowed the Puritan scientists to escape the taint of heresy associated with the occultism and animism of the radical sectaries.

*See glossary of terms.

More important, it allowed the reformers to attack the radicals. The idea that matter is moved mechanically by the will and according to the intelligence of a supernatural God upheld the orthodox Christian dualism of matter and spirit in the face of the radicals' animism, their belief that all matter was endowed with soul and that spirit was immanent in nature.[16] Nor was this merely a victory against false theological doctrine; it also had religious and political ramifications. The vitalistic or pantheistic idea of nature provided the metaphysical grounds for an attack on traditional authority in church and state. If spirit lay within man and nature, this was a strong argument against organized churches, supported by tithes and learned ministries, that claimed superior spiritual wisdom and separate spiritual authority—the power to teach, discipline, and punish.[17] Vitalism, with spirit diffused equally in the material world, could also be used to support the notion of human equality and to justify in cosmic terms antimonarchical and even democratic political ideas. The natural philosophy of the radicals tended to dissolve hierarchy, while hierarchical social order found support in the Christian dualism newly shored up by the corpuscular philosophy of the reforming Puritans.[18]

The inductive or experimental aspect of the new corpuscular philosophy worked out in the 1650s also bore an ideological message meant to counter the radicals. Scientific progress would come through painstaking inquiry, the collection of evidence, and the testing of hypotheses. Knowledge then was not, as the sectaries with their emphasis on magic and the occult maintained, the result of mystical experience or God's direct revelation to the saints.[19] God instead revealed himself indirectly by two means, nature and Scripture, his work and his word, and both required close study in order to bear fruit. This emphasis on patient, industrious scrutiny was directed against the antinomian theology of the radical sectaries, which insisted that God revealed himself immediately to the saints so that they might achieve perfection or at least perfect wisdom in this life. The fruits of salvation were accessible here and now as well as in the world to come. For the Puritan reformers, on the other hand, the effortless pleasures of salvation were deferred to the next life; in this life rewards would come only through reason and industry.[20] Science, the new philosophy, was the model; knowledge would come not through visions or divine illumination but rather through a searching and sustained inquiry into nature, the humility and dedication of the experimental philosopher. Nor was this update on the work ethic only directed against the illuminism of the sectaries; it was also seen as an instrument of social control. As Boyle insisted, hard work would keep men too busy to contrive heresy or to plot social revolution.[21] Science would be particularly valuable in this regard because the practical application of its discoveries would create more and more employment.[22] Both the corpuscularianism and the experimentalism of the reforming philosophers were designed to combat two threats, heresy and social insubordination, at the same time.

The reforming Puritan savants also deployed their corpuscular philosophy against Hobbes and Hobbesists. Indeed after 1660 Hobbesism was increasingly identified with subversion[23]— and with good reason, as we shall see. Corpuscularianism preserved a role for spirit in the universe, namely, that of imparting motion to matter and giving shape to the world through providential design. Hence the corpuscularians, against Hobbes's materialist surgery, supported the imposition of order from above and therefore the clergy's authority as interpreters of God's ways and will.[24] The experimentalism of the Puritan savants also offered the way to knowledge through induction and the testing of hypotheses—not through Hobbes's deductive rationalism based in part on mathematical reasoning, any more than through the saints' illuminism.[25]

Under the impact of radical sectarian and Hobbesist challenges the Puritan philosophers grafted their reforming science onto an ideology that sought to reestablish order and stability in church and state. Science was not only seen to alleviate man's material condition; it might also cure the excesses of revolution. The natural philosophers who created the new ideology of science, such as Boyle and Wilkins, kept the reforming aims of the Puritan scientific vision, particularly when they were easily accommodated to, or even promoted, the larger political and religious goal. So they continued to argue for science as a means to greater private profit and national wealth and power, because science, to the extent that it increased agricultural production, trade, and shipping, would also foster domestic peace. The scientific protagonists equated a science made practical with growing prosperity, social order, and the public good.[26]

THE ANGLICAN ORIGINS OF MODERN SCIENCE

The initial Puritan reforming vision of the 1640s thus survived in a continuing belief in the material benefits of science.[27] By the late 1650s, however, this belief in science as an instrument of material progress was wedded to a new Anglican theology, one no longer essentially Puritan but liberal or latitudinarian. Its central tenets were the repudiation of predestination, a concomitant emphasis on free will and striving as the keys to salvation, and an almost obsessive concern for design, order, and harmony as the primary manifestations of God's role in the universe.[28] Evolved during the 1650s, this liberal Anglicanism relied on the reforming vision of the new science to verify both God's order in an unstable world and the superiority of cautious scientific inquiry over the illuminations of the spirit. During the Cromwellian Protectorate the hope among men like Boyle and Richard Baxter was that this view could be translated into a church settlement based on Archbishop James Ussher's schemes for moderate episcopacy.[29] These hopes, of course, were never realized; but they survived the Restoration, and this scientifically grounded latitudinarianism* received its classic formulations in the works of Robert Boyle published after 1660 and in Thomas Sprat's *History of the Royal Society* (1667), masterminded by John Wilkins. It was thus adopted as the public stance, if not the official ideology, of the Royal Society.[30]

The importance of liberal Anglicanism for the integration of the new science into the mainstream of English and eventually European thought cannot be easily overestimated. It was a scientifically grounded social ideology, as well as a new religious piety, that endorsed experimentalism and material progress based on science in a way in which no other contemporary social or religious vision could do. It made science a fit subject for pulpit discourse and in so doing rendered it more immediate to daily

*See glossary of terms.

thought and experience than did the weighty tomes of the Cartesians. And not least, the new mechanical philosophy as articulated by Boyle and his circle managed to escape the trap set by Cartesian dualism or Hobbesist mechanism; it was resolutely antimaterialist, not to mention antimagical, and hostile to sectarianism. The importance of this late seventeenth-century English synthesis permits us to speak of the Anglican origins of modern science as being not opposed to but superseding the Puritan origins of modern science. Liberal Anglicanism provides the ideological continuity between the science of Boyle—that is, the experimental method of modern science—and the science of Isaac Newton (b. 1642). And with Newton, the new science achieved maturity. The general history of its integration in the eighteenth century, as we shall see in Chapters 5 through 7, centers in good measure on the spread of Newtonian science as an understandable and usable synthesis by which the physical order may be explained and exploited mechanically.

The cultural milieu of the English Revolution brought into sharp relief the social implications of the varieties of human understanding of nature available to early modern Europeans. To capture nature's power magically conferred power on self-proclaimed priests and prophets alike. Hence once again, naturalistic and hermetic doctrines in the hands of ordinary men justified a heady arrogance against established authority. Scholasticism remained the true ally of Catholicism; and for English Protestants, with plenty of evidence to point to on the Continent, Catholicism meant absolutism. In the face of all these godly assaults on that purely secular entity, the state, Hobbes offered a basic materialism as the only sane alternative. Amid the absolutists, radicals, and materialists, landed gentlemen of Protestant persuasion wanted to retain a state religion, but with a clergy docile to them and without the upheavals from the middling and lower orders that the revolution had unleashed. Within this context and in the face of so many unacceptable alternatives, after 1660 Puritanism gave way to liberal Anglicanism, and the mantle of science passed to a new generation of intellectual leaders. Out of that generation came the metaphysical and religious assumptions that made possible the Newtonian synthesis.

But before that extraordinary Newtonian synthesis could be achieved powerful reasons had to be found for a complete repudiation of Cartesianism. Among the Continental proponents of the new science such a total rejection of Descartes seemed unnecessary, if not bizarre. While there were problems clearly evident in aspects of Descartes's cosmology and physics, his unrelenting insistence on mechanisms and contact between bodies seemed the only viable alternative to the occultism of the magicians or the qualities and forms of the scholastics. Consequently, as we have seen (pp. 65–69), Cartesianism made slow but steady progress in the universities of Continental Europe, particularly in Protestant countries, and indeed in England and Scotland. Only in Cambridge in the 1660s among the liberal Anglican opponents of the radical sectarians and Hobbes, did Descartes come to be seen as untenable because of the materialist and nonexperimental implications of his system. The English Revolution and the reaction to it created the ideological climate that in certain circles undermined Cartesian certainty. In that sense we may say that while the culmination of the Scientific Revolution is unthinkable without Newton, Newton is unthinkable without the English Revolution.

If we focus attention very precisely on the Cambridge colleges of the 1660s, on that moment when the young Newton came to Trinity as an undergraduate (1661) and on the extant printed and manuscript evidence available there from that period, we can witness the intellectual revolution to which that young student of natural philosophy was exposed. In the 1650s the main tenets of Puritanism had been firmly repudiated among certain philosophers and college fellows who nevertheless wished to retain the new science. They too had repudiated scholasticism and turned, therefore, to the Christianized Platonic tradition of the Renaissance to search for explanations of nature that would counter Aristotle while preserving the basic tenets of Protestant Christianity and the immediacy of God's presence in his creation.[31] Neo-Platonism, it was believed, when wedded to the new science, would preserve mechanical action while retaining spiritual forces in nature.

The leaders of this Cambridge school were Henry More and Ralph Cudworth. In More's earliest published work, a collection of poems, *Platonica* (1642), he sought to articulate a Platonic sense of spiritual forces in nature that could be understood scientifically. At first he was also powerfully attracted to Descartes's writings and even corresponded with the French philosopher.[32] In this same period More observed with horror the dislocations produced by the civil wars and the interregnum. He came in turn to despise enthusiasm and Puritanism—"such horrid errors, that they seem the badges of the kingdom of darkness"—just as he detested Catholicism (*Divine Dialogues,* 1668).

By 1653, however, More was also voicing reservations about the Cartesian system; in 1665 he registered to Boyle his complete rejection of Descartes based on his fear that Descartes's system, just like those of Hobbes and Epicurus, led directly to atheism. More believed that true atomism required the assertion of spiritual forces in nature and a rejection of a purely mechanical and material causation. In the words of one commentator, that rejection "provided an indispensable background to Newton's theory of active principles and preoccupation with the role of the spiritual in nature."[33] Without that belief in active principles, Newton could never have postulated the existence of universal gravitation as an immaterial force operating throughout the universe independent of any direct, mechanical contact action between bodies.

ISAAC NEWTON AND RESTORATION CAMBRIDGE

In the early 1660s the Cambridge Platonists sought, in their own words, to give the new generation of undergraduates an alternative to both Descartes and Aristotle:

> And seeing that they will never return to the old Philosophy, in fashion when we were young scholars, there will be no way to take them off from idolizing the French Philosophy and hurting themselves and others by some principles there, but by putting into their hands another body of Natural Philosophy, which is like to be the most effectual antidote.[34]

In Newton's earliest student notebook of 1663 we can discern that search for the antidote. He is drawn to the atomism of Gassendi at this early stage, and he is repelled by Descartes's definition of matter as an infinitely extended plenum; the young undergraduate reasons that if all the universe were filled with matter, then there would be no room for motion. Atomism, on the other hand, permits vacuity between the particles, and it became one of the cornerstones of Newton's mature philosophy of nature.[35] From those early student jottings it is evident that Newton has abandoned Aristotle and has encountered Descartes but found the latter (but not his mathematics or the new science per se) wanting. In consequence he embarked on an intellectual odyssey totally dominated by contemporary scientific problems. Those notebooks also reveal that Newton was in touch with natural philosophical matters frequently discussed within select circles at the university. In his notebook he is responding to standard tutorial questions *(quaestiones)*[36] as well as to those philosophical issues. We know that in this period Hobbes and Descartes were being read at the university, while in 1667 the vice chancellor of Cambridge University publicly condemned the reading of Descartes by candidates for the baccalaureate.[37]

In the 1660s, after the restoration of the monarchy, the polemic against Hobbesism, enthusiasm, and naturalism began in earnest. While Boyle and Joseph Glanvill pounded at the naturalists in print, More and Cudworth in Cambridge worked out a variety of attacks against Hobbesism, Cartesianism, enthusiasm, and yet another version of materialism coming from the Dutch Republic in the form of Spinoza's pantheism (see pp. 117–119). Yet it should be noted that many of these same conservative reformers still retained one vital element in the old Puritanism. While repudiating predestination and the "reign of the saints"— that is, those men and women who had sought in the 1650s independence from ecclesiastical authority—the conservative reformers based in Cambridge continued to believe in the possibility of a millenarian paradise. Indeed their millenarianism, now conceived to include no alteration of the existing system of ecclesiastical and political authority, did nevertheless postulate an earthly paradise wherein the righteous would rule.

Newton's private writings from the 1660s echo this same polemical rhetoric. His manuscripts and notebooks from this period when he formulated the metaphysical positions at the foundation of his science, ones that stayed with him until his death, reveal his millenarianism. Most important, they make use of rhetorical formulations of natural philosophy directly relevant to the ideology of the new Anglicanism. Shortly after Newton's death, his associate John Craig, who many years earlier had been the intermediary between Newton and the young Newtonian polemicist Richard Bentley, wrote that the reason for Newton's "showing the error of Cartes' Philosophy, was because he thought it was made on purpose to be the foundation of infidelity."[38]

A close reading of Newton's earliest manuscripts confirms Craig's view. The language he employed was remarkably similar to the Anglican polemics with which he was surrounded. Newton repudiates Descartes's definition of body as extension because it does "manifestly offer a path to Atheism"; likewise he repudiates "the vulgar notion (or rather lack of it) of body . . . in which all the qualities of the bodies are inherent" because it too leads directly to atheism. Newton, like Boyle at the very same time, wanted to construct an alternative to Aristotelian ("vulgar") matter theory because its implications were heretical and specifically because they chimed with the vitalistic and pantheistic notions of "the vulgar" (also another word for the people) that had been spawned by the radical sectaries during the revolution. As Newton says in his manuscript, "Indeed however we cast about we find almost no other reason for atheism than this notion of bodies having, as it were, a complete, absolute and independent reality in themselves." In short Newton saw a profound danger in the specter of atheism, whether in the mechanistic version he read Descartes to be supporting or in the "vulgar" form that denied differences in substance between mind and body, in effect denying "that God exists, and has created bodies in empty space out of nothing." The basic definitions of Newtonian natural philosophy are clearly present in those early manuscripts: the power of divine will to move "brute and stupid" matter; the independent, absolute existence of space and time; and, most essential to the formulation of the concept of universal gravitation, the notion that "force is the causal principle of motion and rest" which operates on bodies *in vacuo.*[39]

This wholesale repudiation of Descartes was essential before Newton could employ his brilliant mathematical skill to formulate precisely the law of universal gravitation. That repudiation began in the 1660s, and it was only late in the 1670s that Newton turned his attention again to the problem of gravity; the full-scale formulation of his law emerged to be published finally in the *Principia* (1687). This is not to say that these religious and ideological factors entirely explain or account for his scientific brilliance. They permitted it to flourish in the direction that it did, to be sure; but at some point the historian must acknowledge the presence of a creative power, particularly in mathematics, of unprecedented force. We can only speculate as to how far religious beliefs and ideological concerns, so particular to the period after 1660, compelled the young Newton to search for evidence of divine efficacy in every aspect of the material order, in effect to develop as a scientist. The religious Newton was never at odds with the scientific Newton; quite the reverse.

For the most part, both impulses were part of his very private world. He chose to publish his science only when pressed; his religiosity lies to this day buried in voluminous private manuscripts. There the historian can glimpse his millenarianism, his hatred for Catholicism, his very liberal ideas on church government, his anti-Trinitarianism (one reason why he remained so private in these matters), and not least, his alchemy. Part of the reason for this "secret" Newton was simply a personal style that was slightly paranoid; part of the reason also relates to the age in which he lived.

The Restoration was a dangerous time for anyone who held to ideas associated with unorthodoxy. Newton practiced alchemy for most of his life; it had once been a cause among the reformers of the 1650s. For Newton, alchemy confirmed his sense of spiritual forces at work everywhere in the universe; indeed these spirits could decompose metals, and once refined, such matter "if it meets a suitable piece, quickly passes into gold."[40] But it would not do to publish such beliefs; they too lay buried in a multitude of alchemical manuscripts.[41]

Even Newton's most singular contributions to the new science stayed buried for a time amid his private and youthful papers. Perhaps the greatest period of his creativity occurred in the mid-1660s, when he discovered the calculus; formulated the inverse square relationship between the sun and the planets, the earth and the moon; and through experimentation with light filtered through a prism, determined that colors are not complex modifications of light but rather each color is unique and possesses its own degree of refrangibility. We can summarize these discoveries of the mid-1660s in Newton's own words:

> In the beginning of the year 1665 I found the Method of approximating series & the Rule for reducing any dignity of any Binomial into such a series. The same year in May I found the method of Tangets of Gregory & Slusius, & in November had the direct method of fluxions & the next year in January had the Theory of Colours & in May following I had entrance into ye inverse method of fluxions. And the same year I began to think of gravity extending to ye orb of the Moon & having found out how to estimate the force with w^{ch} [a] globe revolving within a sphere presses the surface of the sphere from Keplers rule of the periodical times of the Planets being in sesquialterate proportion of their distances from the centres of their Orbs. I deduced that the forces wch keep the Planets in their Orbs must [be] reciprocally as the squares of their distances from the centres about wch they revolve; & thereby compared the force requisite to keep the Moon in her Orb with the force of gravity at the surface of the earth, & found them answer pretty nearly. All this was in the two plague years of 1665 & 1666. For in those days I was in the prime of my age of invention & minded Mathematics & Philosophy more than at any time since.[42]

When Newton was no longer "in the prime of [his] age" he pursued alchemy, theology, and church history with the same avidity that he had once brought to mathematical and natural philosophy. Indeed in the 1680s he, like so many other Anglicans, became once again obsessed with the meaning of the scriptural prophecies, with the final days of the world.

THE REVOLUTION OF 1688–1689 AND THE NEWTONIAN SYNTHESIS

While the Anglican natural philosophers of the Restoration had successfully beaten back the threat to orthodoxy and the hegemony of the church once presented by the radical sectaries,[43] other dangers lurked around every corner. The sophisticated materialism of Hobbes as well as the republican legacy of the 1650s continued to attract elite followers, particularly among the opponents of royal power. But in the 1680s the specter of monarchical absolutism returned in the person of James II (1685–1688), who was also a devout Catholic.

Suddenly the hegemony of the Anglican clergy appeared directly threatened, and all other clerical institutions concerned with education or welfare, such as the Oxford and Cambridge colleges, also felt the cold scrutiny of a king intent on installing Catholics in high places. The Royal Society, like so many of those colleges, had lent its support to royalism during the Restoration; evidence suggests that at various critical moments in the late 1670s and early 1680s Fellows of the Royal Society had written in support of monarchical authority.[44]

James II inherited from his brother, Charles II, a court that was not only absolutist by inclination but also notorious for its private libertinism. Yet it was also open to the intellectual interests of its day. The French Epicurean Saint-Evremond had a following within it, while Charles himself, although largely ignorant of matters philosophical, had offered his protection to the Royal Society.[45]

Within this context established by royal patronage and the fear of political instability, the Royal Society attempted throughout the Restoration to chart its fortunes and those of the new science. It sought, as one modern commentator has put it, "to bring rationalization and order to all areas of national life."[46] Great emphasis was placed on technological improvements, on mechanical devices intended for industry and agriculture. The inspiration for these projects was Baconian, or in some cases the motivation came from direct requests by government agencies for the Society to assist in one or another project.[47]

Individual Fellows, not least among them clergymen like Joseph Glanvill, had direct ties with their rural parishes and the needs and interests of the local gentry. In early modern England the rural economy had come to include industrial development —mining for coal and minerals in particular, but also light manufacturing done by water power. In the records of the Society from the 1680s we find evidence of interest in the earliest steam engines; and most important, the Society was receptive at that early date to what was to become a socially revolutionary argument. The Fellows discussed the notion that mechanical devices should and indeed could save labor, in effect to decrease rather than increase employment.[48] At the time of those discussions it was extremely difficult to get a patent from the government for any device if its inventor argued that it would *save* labor. Indeed until the late 1720s patents may have been rejected if an applicant argued such a case.[49] Yet in the minds of Restoration natural philosophers associated with the Royal Society we can find a mentality discernibly industrial in the modern meaning of that term and, most important, an eagerness to promote their vision of industrial progress whatever the immediate and, from the government's point of view, undesirable social consequences. The alliance forged in the 1660s between the new science and the landed and commercial elite whose interest prospered in the late seventeenth century possesses historical implications that stretch right down to the late eighteenth century and the Industrial Revolution (see Chapter 5).

Having allied itself with the search for order, stability, and the growth of commercial and industrial enterprise, the Royal Society was dependent on monarchical support for its continuing respectability. It is little wonder then that in the late 1680s, when James II's absolutist policies threatened to destabilize the political order, undermine the Anglican church, and plunge the country into a new civil war, the Royal Society cast about to secure its interests as well as to remind the new king of his obligations.

At just that moment (in 1687) the *Principia* of Isaac Newton was published under the imprimatur of the Society. This is a singularly important date in the history of Western thought. After it we are able to speak of the existence of a Newtonian synthesis, an ensemble of scientific laws—specifically, the law of universal gravitation, proven mathematically and in turn capable of being demonstrated experimentally by the use of mechanical devices—a particular natural philosophy, neo-Platonic in origin and antimaterialist in intention, a polemically hostile anti-Cartesianism, and not least, a series of social and political explications to be drawn from this Newtonian model of cosmic order. As we have seen, the emergence of that synthesis bears relation to the ideological struggles we associate with the English Revolution. And not least, the timing of the publication of the *Principia* may also owe something to those political uncertainties that were only to be resolved by the flight and constitutional expulsion in 1688–1689 of James II.

In the light of what is now known about the political activity of the Royal Society during the Restoration, we should at least consider the question of why the *Principia* appeared when it did. The standard story is that Edmond Halley, a Fellow of the Royal Society and a friend to Newton, prodded the reticent and otherwise preoccupied Newton into writing and publishing his magnum opus. Throughout the instabilities of the 1680s, it should be remembered, Newton seems to have been particularly preoccupied with the rise and fall of ancient monarchies and with the apocalyptic texts of the Old and New Testaments.[50] But Halley persuaded him to leave his historical and alchemical studies when he brought news about debates in London on the phenomenon of universal gravitation. The result of that digression was, of course, the *Principia.* It bore on its title pages the imprimatur of the Royal Society as well as the name of Samuel Pepys in bold type. Pepys in this period was avidly seeking favor at the court of James II, and indeed he paid dearly for his sycophancy in the postrevolution wilderness after 1688–1689 to which he was consigned.

The difficulty with the Halley-Newton story, attractive though it is, is that it also closely resembles George Ent's description of his role in prodding William Harvey to allow his *De generatione animalium* to be published in 1651.[51] That, of course, does not make the story untrue in the case of Newton. But if there is something to the hypothesis that the publication of Newton's *Principia* during the reign of James II was inspired by political motives, to which Newton may very well not have been privy, we would expect some oblique indication of that covert design, some hint dropped—either in Halley's admiring ode on Newton and his achievement prefixed to the *Principia* or possibly in Halley's fawning and explanatory letter on that achievement addressed to James II and later published in the *Philosophical Transactions.*[52]

A new translation of Halley's ode, which relates it directly to Lucretius's *De rerum natura,* has been most usefully supplied by W. R. Albury. That ode begins by reminding the *Principia*'s readers that "the pattern of the Heavens" is based on "Laws which the all-producing Creator, when he was fashioning the first-beginnings of things, wished not to violate and established as the foundations of his eternal work." After this brief mention of the eternality of "Law" and the role of the "Divine Monarch" as its creator and preserver, the poem goes on to glory in the power

unleashed by Newton's intellect, which "has allowed us to penetrate the dwellings of the Gods and to scale the heights of
Heaven." Couched entirely in Epicurean language, Halley's ode
commends the new science sponsored by the Royal Society as the
means by which "we are truly admitted as tableguests of the
gods."[53] We would agree with Albury's conclusion that the explanation for Halley's choice of Lucretius as his model must lie in
his attempt to "reply to Epicurean criticism of the Royal Society"
which had circulated in court circles during the 1680s. In short,
Halley is trying "to convince the fashionable Epicureans associated with the Stuart Court of the importance of Newtonian
science,"[54] and this at an absolutely critical time for Anglican
natural philosophers and churchmen, who had been systematically excluded from James's court.

Perhaps we can now better understand why, after the church's
hegemony had been reestablished in the early 1690s, Newton
wrote a seemingly bizarre letter to Pepys in which he asserts,
almost hysterically, "I never designed to get anything by your
interest, nor by King James's favour."[55] If indeed the *Principia*
had been published in an attempt to be ingratiating, by reestab-
lishing the supportive role that scientific knowledge had given
the monarchy during the Restoration, then either Newton was
innocent of these motives or he became almost paranoid after the
revolution in his concern that his name not be associated with
Pepys, who was by that time suspected of Jacobitism (that is, of
still supporting James II).

If Newton was in all probability naive in 1686, he was not so
by 1692. He had led the anti-Catholic opposition to James II in
Cambridge and had given his assent to the Revolution of 1688–
1689. He urged his parliamentary constituents to do likewise.[56]
But the revolution undid more than the Stuarts. It secured the
church's constitutional place yet vastly weakened its legal and
moral authority. The latitudinarian faction had now ascended to
positions of leadership within the church's hierarchy, and its
problems had become theirs.

In the period after 1689 Newton's natural philosophy served
as the underpinning for the social ideology preached by the
church in response to the Revolution Settlement. The Newtonians once again resumed the polemical assault against intellectual
and political radicalism, and they did so in language characteris-

tic of Restoration Anglican science. They spoke, fittingly, from the podium established by Boyle's last will and testament (1691). With Newton's assistance and approval the Boyle lecturers—Richard Bentley, Samuel Clarke, William Whiston, and William Derham—brought Newton's "system of the world" to bear against the radical Whigs* of the 1690s and beyond, those whose heterodox religion of nature owed much to their reading of Hobbes and Spinoza as well as to Bruno and Servetus, to the extreme pagan naturalism of the late Renaissance.[57] Indeed the Boyle lecturers did precisely what Newton had indicated to a friend in late 1691 could to be done: "A good design of a publick speech (and which may serve well as an Act) may be to shew that the most simple laws of nature are observed in the structure of a great part of the Universe, that the philosophy ought there to begin."[58]

From Boyle's endowed pulpit and in their writings the Newtonians preached to London-based and exceedingly prosperous congregations. They extolled the virtues of self-restraint and public-mindedness while at the same time assuring their congregations that prosperity comes to the virtuous and that providence permits, even fosters, material rewards. Men must acknowledge God's providence by the cultivation of virtue, by the pursuit of what Newton's tutor, Isaac Barrow, had called "sober self-interest," and by their support for Anglican hegemony. The same God whose laws of motion Newton had discerned in the natural world would also inevitably ensure order, prosperity, and the conquest and maintenance of empire in the political world. Adopting the language of the scientific novice, Newton's advocates used his science, as Restoration Anglicans had used theirs, to support the social ideology and political goals of the liberal Anglicanism that had been rendered supreme within the recently secured church.

With the lapse of the Licensing Act in 1695 and the intensification of party politics in the late 1690s, the liberal Anglican establishment, along with court and monarchy, found itself under assault by the radical and republican Whigs.[59] With the free-thinker John Toland in the vanguard, they put forth materialistic and pantheistic—to use the word invented by Toland in 1705—arguments to justify the rule of Parliament over court placemen

*See glossary of terms.

and standing armies, of civic religion over the established church, of religious pluralism over a narrowly circumscribed toleration. From their pulpits the Boyle lecturers, with the Newtonian Samuel Clarke as their most philosophically gifted spokesman,[60] put forth arguments to justify order and stability, to maintain the hierarchical and providential interpretation of the constitutional settlement.

But if Anglican hegemony now owed so much to Newtonian science, what did Newton's science owe to its religious and ideological roots? On the crucial level of matter theory, on Newton's insistence that universal gravitation must operate through immaterial forces in the universe and not as a property inherent in matter, it seems plausible to argue that Newton had accepted the central arguments of the Anglican virtuosi as formulated during the 1650s and beyond. Certainly his private manuscripts from as late as the 1690s repudiated the materialistic arguments by which "the vulgar" described the world and lashed out at those who postulated an impotent deity—a "dwarf-god," as Newton put it.[61] Newton's insistence on a mechanical philosophy that relied heavily on spiritual forces led him to adopt a baroque and neo-Platonic ontology that to this day has puzzled those purely philosophical commentators intent on unraveling its complexity.[62] This approach does not seek to minimize that complexity, but it does offer one explanation for its existence.

If we date the origins of the European Enlightenment to the 1690s in England, then it is now increasingly clear that English science from Boyle to Newton sponsored but one version of Enlightenment. Given what we now know about the institutional and ideological relations of the new science, in short about its Anglican origins, it must be acknowledged that, far from "preparing the ground for the deists of the Enlightenment,"[63] the Newtonian Enlightenment was intended by its participants as a vast holding action against materialism and its concomitant republicanism, against what is best described as the Radical Enlightenment.

This Newtonian Enlightenment cast its light in a variety of directions. From England it spread quickly onto the Continent, largely through the assistance of the French language press at work in the Dutch republic. There, as in England, its advocates first attacked the Cartesians; then, once established as legitimate,

Newtonian science, particularly in its mechanical application, appealed to industrial promoters as well as to philosophers and social reformers. The model of order based on knowable laws embodied in the Newtonian synthesis offered a powerful alternative to a variety of other belief systems, not least to the doctrines of the scientifically naive clergy. With the dissemination of the new science in the early eighteenth century, the break between high culture and low culture was now complete. For those European elite who also embraced science the goal became enlightenment, and England and its science became the model of order, stability, and progress.

NOTES

1. Christopher Hill, *The Century of Revolution 1603–1714* (London: Nelson, 1961); the articles by S. F. Mason, H. F. Kearney, Christopher Hill, T. K. Rabb, Barbara Shapiro, and Margaret 'Espinasse that first appeared in *Past and Present,* collected since in Charles Webster, ed., *The Intellectual Revolution of the Seventeenth Century* (London and Boston: Routledge and Kegan Paul, 1974), pp. 197–316, 347–368; P. M. Rattansi, "The Social Interpretation of Science in the Seventeenth Century," in Peter Mathias, ed., *Science and Society 1600–1900* (Cambridge: Cambridge University Press, 1972), pp. 1–32; Margaret C. Jacob, *The Newtonians and the English Revolution 1689–1720* (Ithaca, N.Y.: Cornell University Press, 1976); and J. R. Jacob, *Robert Boyle and the English Revolution* (New York: Burt Franklin, 1977). Portions of this chapter first appeared in James R. Jacob and Margaret C. Jacob, "The Anglican Origins of Modern Science: The Metaphysical Foundations of the Whig Constitution," *Isis,* vol. 71 (1980), pp. 251–267.

2. Robert K. Merton, *Science, Technology and Society in Seventeenth-Century England* (New York: Howard Fertig, 1970); and Charles Webster, *The Great Instauration: Science, Medicine and Reform, 1620–60* (London: Duckworth, 1975). Note the perceptive comments by Benjamin Nelson in Tom Bottomore et al., eds., *Varieties of Political Expression in Sociology* (Chicago: University of Chicago Press, 1972), pp. 202–210. For a recent defense of the Merton thesis, unreformed, see Gary A. Abraham, "Misunderstanding the Merton Thesis: A Boundary Dispute Between History and Sociology," *Isis,* vol. 74 (1983), pp. 368–387.

3. H. A. M. Snelders, "Science in the Low Countries During the 16th Century: A Survey," *Janus,* vol. 70 (1983), pp. 213–227; the great exodus of intellectuals out of the southern Netherlands after the

Spanish conquest in 1585 had "a paralyzing effect on the culture" of that region. For the teaching of science in a typical Dutch academy of the seventeenth century, see Rijksarchief, Gelderland, MSS, Academie to Harderwij, no. 154. Catalogue of the library includes Gassendi and Bacon in 1671; Descartes and John Ray by 1698. Cf. Th. J. Meijer, "De historische achtergronden van wetenschappelijk onderzoek in Leids universitair verband," *Tijdschrift voor geschiedenis,* vol. 85 (1972), pp. 432–443.

4. Webster, *The Great Instauration,* pp. 90–96, 259. For the role of religion in the revolution, without accepting its conclusions, see John Morrill, "The Religious Context of the English Civil War," *Transactions of the Royal Historical Society,* 5th ser., vol. 34 (1984), pp. 155–178.

5. P. M. Rattansi, "Paracelsus and the Puritan Revolution," *Ambix,* vol. 11 (1963), pp. 24–32.

6. Ibid., p. 27. Cf. Thomas H. Jobe, "The Devil in Restoration Science: The Glanvill-Webster Witchcraft Debate," *Isis,* vol. 72 (1981), pp. 343–356.

7. Christopher Hill, *The World Turned Upside Down* (London: Temple Smith, 1972), Chapter 14.

8. Robert Boyle, *Some Considerations Touching the Usefulness of Experimental Natural Philosophy* (London), Part 1 (1663) and Part 2 (1671). Both parts were written during the 1650s; see R. S. Westfall, "Unpublished Boyle Papers Relating to Scientific Method," *Annals of Science,* vol. 12 (1956), p. 65; and Thomas Birch, ed., *The Works of the Honourable Robert Boyle,* 6 vols. (London, 1972), vol. 3, p. 395. For a treatment of Parts 1 and 2, see James R. Jacob, *Boyle,* pp. 104–118 and 141–143, respectively. See also Charles Webster, "The College of Physicians: 'Solomon's House' in Commonwealth England," *Bulletin of the History of Medicine,* vol. 41 (1967), pp. 393–412; J. J. O'Brien, "Commonwealth Schemes for the Advancement of Learning," *British Journal of Educational Studies,* vol. 16 (1968), pp. 30–42; and Christopher Wren, *Parentalia: Or Memoirs of the Family of Wrens* (London, 1950), p. 196.

9. J. R. Jacob, *Boyle,* pp. 141–143; and Royal Society of London, Letter Book Supplement, A-B Copy, John Beale, pp. 348, 382, 389–390, 403–410.

10. Allen G. Debus, ed., *Science and Education in the Seventeenth Century: The Webster-Ward Debate* (London: Macdonald, 1970); Henry Stubbe, *A Light Shining out of Darkness* (London, 1659), which was "answered by H. F. [Henry Ferne?] but never printed," according to Anthony á Wood, *The History and Antiquities of the University of Oxford,* 3 vols. (Oxford, 1792–1796), vol. 3, p. 695; for the conservative reaction to Stubbe's attack on conventional religion and the universities: Anthony á Wood, *Athenae Oxoniensis,* ed. P. Bliss, 4 vols. (London, 1813–1820), vol. 3, p. 1069. See also: *Sundry Things from Several Hands Concerning the University of Oxford* (London, 1659); and Charles

Webster, "William Dell and the Idea of University," in Mikulas Teich and Robert Young, eds., *Changing Perspectives in the History of Science* (London: Heinemann, 1973), pp. 110–126.

11. J. R. Jacob and M. C. Jacob, "Scientists and Society: The Saints Preserved," *Journal of European Studies*, vol. 1 (1971), pp. 87–90; and J. R. Jacob and M. C. Jacob, "Seventeenth Century Science and Religion: The State of the Argument," *History of Science*, vol. 14 (1976), pp. 200–201.

12. Quoted in M. C. Jacob, *The Newtonians*, p. 14.

13. Christopher Hill, *The Religion of Gerrard Winstanley*, supplement 5, Past and Present Society, 1978, p. 18.

14. Ibid. For a good general introduction to Winstanley, see G. E. Aylmer, "The Religion of Gerrard Winstanley," in J. F. McGregor and B. Reay, eds., *Radical Religion in the English Revolution* (Oxford: Oxford University Press, 1984), pp. 91–120. For the beliefs of "ordinary folk" in this period, see also Margaret Spufford, *Small Books and Pleasant Histories: Popular Fiction and Its Readership in Seventeenth-Century England* (London: Methuen, 1981).

15. Marie Boas, *Robert Boyle and Seventeenth-Century Chemistry* (Cambridge: Cambridge University Press, 1958); Robert H. Kargon, *Atomism in England from Harriot to Newton* (Oxford: Clarendon Press, 1966), pp. 93–105. Cf. Steven Shapin and Simon Schaffer, *Leviathan and the Air-Pump: Hobbes, Boyle, and the Experimental Life* (Princeton: Princeton University Press, 1986).

16. J. R. Jacob, *Boyle*, pp. 112–115.

17. Thomas Edwards, *Gangraena*, 3rd ed. (London, 1646), Division 1, Part 1, pp. 25–26; Hill, *The World Turned Upside Down*, Chapter 6.

18. Edwards, *Gangraena*, pp. 15–19, 23–24, 28–29; J. R. Jacob, *Boyle*, Chapters 3 and 4; and M. C. Jacob, *The Newtonians*, Chapter 1.

19. J. R. Jacob, *Boyle*, pp. 98–112.

20. Ibid., pp. 85–88.

21. Peter Pett, *A Discourse Concerning Liberty of Conscience* (London, 1661), p. 9. This was a tract commissioned by Boyle and representative of his views (J. R. Jacob, *Boyle*, p. 134).

22. Thomas Sprat, *A History of the Royal Society* (London, 1667), pp. 343, 400, 408, 425–429.

23. [Thomas Tenison], "The Epistle Dedicatory," in *The Creed of Mr. Hobbes Examined* (London, 1671), pp. 7–8, 13–15; Joseph Glanvill, *A Blow at Modern Sadducism* (London, 1668), pp. 153–160; John Evelyn, *The History of Religion*, ed. R. M. Evanson, 2 vols. (London, 1850), vol. 1, pp. xxvii–xxviii; and J. R. Jacob, "Civil Religion and Radical Politics: Stubbe to Blount," paper presented at the annual meeting of the American Historical Association, San Francisco, 1978.

24. John Wallis to John Owen, October 10, 1665, in Peter Toon, ed., *The Correspondence of John Owen (1616–1683)* (Cambridge: Cam-

bridge University Press, 1970), pp. 87–88; John Wallis, *Hobbiani Puncti Dispunctio* (Oxford, 1657), pp. 42–43; and Robert Boyle, "The Preface," in *An Examen of Mr. T. Hobbes His Dialogus Physicus de Natura Aeris* (Oxford, 1662).

25. Boyle, "The Preface," *An Examen;* and John Wallis, "The Epistle Dedicatory," in *Elenchus Geometriae Hobbianae* (Oxford, 1655).

26. Wren, *Parentalia,* p. 196; and Royal Society, Letter Book Supplement, A-B Copy, John Beale, pp. 348, 382, 389–390, 403, 410. For a suggestive approach to politics and economic ideology in this period, see Joyce Appleby, *Economic Thought and Ideology in Seventeenth Century England* (Princeton: Princeton University Press, 1978), Chapter 9.

27. J. R. Jacob, *Boyle,* pp. 133–159.

28. See Robert Martin Krap, *Liberal Anglicanism: 1636–1647* (Ridgefield, Conn.: Acorn Press, 1944); John F. H. New, *Anglican and Puritan: The Basis of Their Opposition, 1558–1640* (Stanford and London: Stanford University Press, 1964), pp. 16–21. For further evidence of preaching against predestination in Cambridge during the 1650s, see Spencer Research Library, University of Kansas, diary of Charles North, MS A.41, fol. 1, Dr. Cudworth of Clare Hall "On 4 Esiah: 5"; also Dr. Arrowsmith and Dr. Love on the theme "faith without good works is dead." Cf. Gregory Memorandum, Gregory MSS, Edinburgh University Library, DC. 1.61, fol. 93; "When Dr Duport resigned the chair of Greek he recommended his pupil Mr. Barrow who . . . being suspected of Arminianism he could not obtain it and therefore in 1654 he . . . went first to France, in Paris he found his father attending the English Court."

29. G. R. Abernathy, "Richard Baxter and the Cromwellian Church," *Huntington Library Quarterly,* vol. 24 (1961), pp. 227–231; and J. R. Jacob, *Boyle,* pp. 118–126.

30. J. R. Jacob, *Boyle,* pp. 133–143, 152–159.

31. For a discussion of the arguments put forward by the Catholic opponents of the new science, see Edward Grant, "In Defense of the Earth's Centrality and Immobility: Scholastic Reaction to Copernicanism in the Seventeenth Century," in *Transactions of the American Philosophical Society,* vol. 74, part 4 (1984), p. 11 ff.

32. For a valuable discussion of this phenomenon and a review of the historiography, see C. Webster, "Henry More and Descartes: Some New Sources," *British Journal of the History of Science,* vol. 4, no. 16 (1969), pp. 359–377.

33. Ibid., p. 376. Cf. Henry More, *Enchiridion Metaphysicum* (London, 1671).

34. See John Gascoigne, " 'The Holy Alliance': The Rise and Diffusion of Newtonian Natural Philosophy and Latitudinarian Theology Within Cambridge from the Restoration to . . . George III," (Ph.D. diss., Cambridge University, 1981), p. 132; and Gascoigne, "The

Universities and the Scientific Revolution: The Case of Newton and Restoration Cambridge," *History of Science,* vol. 23 (1985), pp. 391–434.

35. For a good exposition of the notebook, see Gale E. Christianson, *In the Presence of the Creator: Isaac Newton and His Times* (New York: Free Press, 1984), pp. 55–56.
36. University Library, Cambridge; Student notebook of John Smyth of Gonville and Caius in 1681, fol. 34 ff., on physics according to Descartes; University Library, Cambridge, MS 6160 notebook of William Bright, November 1645, e.g., ff. 170–176 very similar to Newton's notes; these on God's power and prudence in the government of the world; Add. Mss. 6986 "Dr. Duport's Rules to Fellow Commoners," fol. 9, "When you are ye respondent evermore repeat ye syllogisme before you answer. . . . Write yr. logical and Philosophical rules, distinctions or questions in a little paper pocket book you may carry them about with you." And when the practice became formalized, and incidentally used to teach Newton's science, see *Quaestiones philosophicae in usum juventutis academicae* (Cambridge, 1732); and finally A. R. Hall, "Sir Isaac Newton's Note-Book, 1661–65," *Cambridge Historical Journal,* vol. 9, 1948, pp. 245–250.
37. Bodleian Library, Oxford, MS Rawlinson c. 146, fol. 132.37. Cf. John Gascoigne, "Politics, Patronage and Newtonianism: The Cambridge Example," *Historical Journal,* vol. 27 (1984), pp. 1–24.
38. John Craig to John Conduitt, April 7, 1727, Cambridge University Library, MSS. Add. 4007, fol. 686.
39. "De Gravitatione et aequipondo fluidorum," in A. Rupert Hall and Marie Boas Hall, eds., *Unpublished Scientific Papers of Isaac Newton* (Cambridge: Cambridge University Press, 1962), pp. 142–144, 148. For the singularly important role of this repudiation of Descartes for the development of Newton's natural philosophy, see Richard Westfall, *Never at Rest: A Biography of Isaac Newton* (Cambridge: Cambridge University Press, 1980), p. 381.
40. Burndy Library, Burndy MS 16, fol. 6, r–v.
41. On Newton's alchemy, see Betty Jo Dobbs, *The Foundations of Newton's Alchemy* (Cambridge: Cambridge University Press, 1975); see p. 80 for Newton's link to Hartlibian circles.
42. University Library, Cambridge, Add. MS 3968.41, fol. 85r.
43. See Christopher Hill, *The Experience of Defeat: Milton and Some Contemporaries* (New York: Viking, 1984).
44. J. R. Jacob, "Restoration Ideologies and the Royal Society," *History of Science,* v. 18, February 1980, p. 18.
45. J. R. Jacob, *Stubbe,* p. 107.
46. Michael Hunter, *Science and Society in Restoration England* (Cambridge: Cambridge University Press, 1981), p. 117.
47. Ibid., p. 93.
48. Royal Society MSS C.P. 18, item 8, fols. 66–80.
49. On getting a patent, see Christine MacLeod, "Patents for Invention

and Technical Change in England, 1660–1753" (Ph.D. diss., Cambridge University, 1982), p. 247. Cf. Alan Smith, "Steam and the City: The Committee of Proprietors of the Invention for Raising Water by Fire, 1715–35," *Transactions of the the Newcomen Society*, vol. 49 (1977–1978), pp. 5–18.

50. Frank E. Manuel, *The Religion of Isaac Newton: The Fremantle Lectures, 1973* (Oxford: Clarendon Press, 1974), pp. 99–100; for portions of Yahuda MS 1 by Newton, see Appendix; Manuel, *Isaac Newton, Historian* (Cambridge: Cambridge University Press, 1963), pp. 1–17.

51. Christopher Hill, "Sir Isaac Newton and His Society," in his *Change and Continuity in 17th Century England* (London: Weidenfeld and Nicholson, 1974), p. 274; cf. George Grinnell, "Newton's *Principia* as Whig Propaganda," in Paul Fritz and David Williams, eds., *City and Society in the 18th Century* (Toronto: Hakkert, 1973), pp. 181–192, which at least raises the issue of political motives, although I do not agree with Grinnell's conclusions.

52. On Halley and James II, see I. Bernard Cohen and Robert E. Schofield, eds., *Isaac Newton's Papers and Letters on Natural Philosophy* (Cambridge, Mass.: Harvard University Press, 1958), pp. 397–424; on Halley and Tillotson, see British Library, MSS Add. 17017, fols. 143, 145–146; MSS Add. 4236, fols. 230, 233, 227.

53. W. R. Albury, "Halley's Ode on the *Principia* of Newton and the Epicurean Revival in England," *Journal of the History of Ideas*, vol. 39 (1978), p. 27.

54. Ibid., pp. 36–37.

55. H. W. Turnbull, ed., *The Correspondence of Isaac Newton*, 7 vols. (Cambridge: Cambridge University Press, 1961), vol. 3, p. 279.

56. Ibid., pp. 12–13; cf. Millicent B. Rex, *University Representation in England, 1604–1690* (London: Allen and Unwin), 1954. For Newton's strong interest in his seat, see A. Rupert Hall and Laura Tilling, eds., *The Correspondence of Isaac Newton*, 7 vols. (Cambridge: Cambridge University Press, 1977), vol. 7 pp. 436–437.

57. Margaret Jacob, "Newtonianism and the Origins of the Enlightenment: A Reassessment," *Eighteenth Century Studies*, vol. 11 (1977), pp. 1–25; on the circulation of Servetus manuscripts, see Lambeth Palace Library, MS 933, fol. 74; cf. M. C. Jacob, *The Newtonians*, p. 221.

58. Memorandum by David Gregory, December 28, 1691, Turnbull, *Correspondence of Newton*, vol. 3, p. 191.

59. M. C. Jacob, *The Newtonians*, Chapter 6.

60. Samuel Clarke, *A Demonstration of the Being and Attributes of God: More Particularly in answer to Mr. Hobbes, Spinoza, and their Followers* (London, 1705); cf. John Toland, *Socinianism truly Stated: Being An Example of fair Dealing in all Theological Controversys . . . by a Pantheist to an Orthodox Friend* (London, 1705); Giancarlo Carabelli, *Tolandiana* (Florence: La Nuova Italia, 1975), pp. 119–120.

61. J. E. McGuire, "Existence, Actuality and Necessity: Newton on Space and Time," *Annals of Science*, vol. 35 (1978), p. 470; on More

and Newton as revealed in "De Gravitatione," pp. 471, 480–482; on Spinoza, p. 493. The quotation is derived from J. E. McGuire, "Newton on Place, Time and God: An Unpublished Source," *British Journal for the History of Science,* vol. 11 (1978), pp. 114–123, quoting from Cambridge University Library, MSS ADD. 3965, section 13, fols. 445r–446r.

62. For an excellent illustration of such complexity, see J. E. McGuire, "Neoplatonism and Active Principles: Newton and the *Corpus Hermeticum,*" in Robert S. Westman and J. E. McGuire, eds., *Hermeticism and the Scientific Revolution* (Los Angeles; William Andrews Clark Memorial Library, University of California, 1977), pp. 120–125.

63. P. M. Heimann, "Science and the English Enlightenment," *History of Science,* vol. 16 (1978), pp. 143–151; cf. R. S. Westfall, *Science and Religion in Seventeenth Century England* (New Haven: Yale University Press, 1958), p. 219.

CHAPTER 4

Crisis and Resolution: The Newtonian Enlightenment

The cultural ascendancy of science in the late seventeenth century—from a body of knowledge once promoted by its select devotees in Florence, Paris, or London, to the cornerstone of progressive thought among the educated laity—occurred with extraordinary rapidity. We can date that transformation in the role of science in Western culture quite precisely from the 1680s to the 1720s. Within one generation the transformation was complete in northern and western Europe, and mechanically based science had left the hands of its first crusaders and visionaries and gone into the everyday language of journalists, learned societies, coffeehouse lectures, and church sermons. As a result, it transformed the way urban merchants, progressive aristocrats, and just simply literate gentlemen, and some gentlewomen, understood the physical world around them.

Indeed this assimilation was so rapid, and its impact so great, that historians since the 1930s have identified that period in European culture from the 1680s to the 1720s as one of profound crisis. Out of that crisis emerged a mentality discernibly modern, a new cultural movement, the "age of enlightenment." At that moment, high culture distinguished itself completely and irrevocably from the culture of the people. Science became essential to educated discourse, and so too did heterodoxy. A new rationalism, "active, zealous and intrepid" as one leading cultural historian of the period has described it,[1] became a weapon against Christian orthodoxy and piety, as well as against established authority. Among the educated, new religious persuasions appeared, some indebted to scientific renderings of nature, others derived from popular or ancient sources: Socinianism, Jansenism, deism, pantheism, freethinking. The first denied the Trinity and argued for a universal natural religion; the second (more

105

pious than any of the others) called for a spiritual and pietistic renewal of Catholicism that severed its linkage to monarchical absolutism; the third relegated God to the role of a clockmaker. Pantheism and freethinking—words first used in the early 1700's —amounted to a disdain for all forms of organized religion that led some bold spirits to proclaim nature as the only God. All amounted to a massive shift away from religious toward secular culture or, in the case of Jansenism, toward a critical posture hostile to the traditional support given to monarchy by the French church.

In making that shift possible the new science proved to be a decisive element. It offered a new definition of creation, and hence of the Creator, that fostered a new religiosity. But in the hands of freethinkers science also permitted the first articulation of a coherent universe without any creator. The roots of our uniquely modern ability to examine nature and society as self-contained entities and to offer explanations totally natural, that is entirely human, lie in the crisis of the late seventeenth century. The maturation of that new ability can be found in the naturalistic uses to which science and natural philosophy were put during the eighteenth-century Enlightenment.

In consequence, so many of our contemporary beliefs about science have been inherited from the Enlightenment: a faith in its progressive nature leading to constant improvement of the human condition; its supposed superiority to mere beliefs, opinions, and subjective judgment; the heroic role of the scientist; the presumed need for all other disciplines, however social their focus, to be "scientific"; and not least, the absolute right of free scientific inquiry as an extension of freedom from censorship— a freedom demanded regardless of the social or moral consequences of that inquiry. With this sort of cultural legacy, it is extremely difficult to use our historical imagination and realize that these assumptions emerged as dominant in our culture only early in the eighteenth century. Their rapid acceptance in northern and western Europe was provoked by a European-wide cultural crisis primarily political, yet also social, in its origins.

At that critical moment the new science, whether in its Cartesian or Newtonian form, if coupled with a tolerant version of Christianity, seemed the only alternative to the cultural rigidity and intolerance increasingly associated with the absolutist state and its clergy. This synthesis of science and Christianity also served to combat new varieties of cultural radicalism, the old naturalism of the people, the new naturalism of the literate proponents of materialism and pantheism, and the sectarian enthusiasm of popular religiosity. Such a cultural role for science owed much to the ideological struggles embraced by the seventeenth-century prophets and promoters of science—from Galileo through Gassendi, Descartes, Boyle, and the Cambridge Platonists. When early in the seventeenth century they began to enlist the mechanical philosophy against the culture of "the vulgar," they imagined that some sort of alliance between the established church and state on the one hand and science on the other was possible, indeed inevitable. By the 1680s such an alliance, with the freedom of inquiry and practical application for science that it implied, seemed doomed by the pretensions of absolutist monarchy throughout Continental Europe but now especially in France and England.

THE THREAT OF ABSOLUTISM

In 1685 Louis XIV revoked the Edict of Nantes and sent over one hundred thousand French Protestants into exile in search of religious toleration. Those who remained either converted to Catholicism or faced persecution. Simultaneously Louis XIV embarked on an aggressive foreign policy that threatened the territorial integrity of the Low Countries, both north and south, as well as the western German cities and principalities. In England James II, as we have seen (pp. 91–93), sought to install Catholics within the army and the universities, that is, to undermine the Anglican church. By this time rigid censorship was already an established fact of life in much of Catholic Europe,[2] as was clerical control over the universities. Indeed the absolutist nature of the Hapsburg monarchy in Spain was so long established that its relative decline was largely missed by many astute and hostile observers. Suddenly the 1680s in western Europe resembled the 1580s. Continental Protestants feared for their survival, monarchs once again sanctioned religious persecution, religious refugees crowded into the urban centers of the Low Countries, and Protestant intellectuals went in pilgrimage to London, Amsterdam, and Geneva. Within that context came the crisis out of which the Enlightenment emerged.

Predictably, given the immediate political causes of the crisis, its impact came first in the area of political beliefs and values. Beginning in the 1680s we see a rapid disintegration of confidence in the doctrine of the divine right of kings, an increasing emphasis among political theorists on the rights of subjects rather than on their duties. To justify those attacks on absolutism, natural law theorists of a previous era (such as Hugo Grotius) were invoked, and arguments for the rule of law rather than will became fashionable. The prevalency of such arguments among the political opponents of absolutism may indeed have encouraged a predisposition to the new science among them, a desire for the experimental rather than the simply memorized or doctrinal, and a sympathy with general theories operating according to predictable laws and not whimsical forces. The assault on absolutism and Catholicism also gave credibility to the rhetoric of probability over that of absolute certainty. It is hardly accidental that one of the most subtle historians of Protestantism, the Anglican polemicist Bishop Gilbert Burnet, argued for the probable certainty of knowledge, scientific and otherwise, against the claims to eternal, and hence absolute, authority made by Catholic historians.[3]

The crisis that began in the 1680s also gave a European significance to events and ideologies that were fresh in the minds of English Protestants such as Burnet, Boyle, and their associates. In short, the political crisis of the late seventeenth century brought the legacy of the first of the great modern revolutions into the mainstream of European thought. There it merged with indigenously Continental traditions of anticlericalism, philosophical heresy, and antiabsolutism. The English Revolution of mid-century had produced a body of political, religious, and scientific thought so rich and complex that once discovered by the European opponents of absolutism it became a major source for that synthesis we describe as enlightened. The English version of the new science, whether in the form of Hobbes's materialism or Boyle's Christian atomism, had been, as we have seen in the previous chapter, inextricably bound up since the early 1650s with the search for an alternative to rigid Puritanism, to radical sectarianism, as well as to the pretensions of absolutist monarchy supported by an independent and culturally dominant clergy. The science and natural philosophy of Boyle and Newton in that sense was the by-product of a revolution against the established clergy and the absolutist state. It is little wonder that English theorists—from the liberal Anglican promoters of science to Hobbes and the republicans—appealed to a Continental audience that included French Huguenots,* Dutch lawyers and doctors, French poets of minor aristocratic background (like Voltaire), and an entire generation of Protestant refugee journalists. Exiled in the Dutch republic and armed with their native command of French, these journalists used the freedom of their presses in a vast campaign against absolutism.[4] Not incidentally, through translations and explications they introduced educated Europeans, literate in French, to English science and culture.

THE FAILURE OF THE OLD LEARNING

Coupled with the political origins of the crisis were other cultural factors, by the late seventeenth century more cumulative than traumatic in their effect. The increase in European traffic to non-Western nations had produced a travel literature richly descriptive of customs and beliefs that were totally non-Christian, yet "curiously" moral. Although much racism and Christian chauvinism were mixed into the Western response to the non-West, by the late seventeenth century the cumulative effect of that literature had been to call into question the absolute validity of religious customs long regarded, especially by the clergy, as paramount. And not least, a century of Protestant versus Catholic polemics about the biblical authorization for either version of Christianity had, willy-nilly, rendered the Bible into a historical document. Once reduced to human scale, its contents were open to sceptical scrutiny. Such scrutiny when offered to the literate could only render the task of instruction more difficult for the

*See glossary of terms.

clergy. Simultaneously, literacy increased in England and Scotland (and possibly in the Dutch republic), especially during the second half of the seventeenth century. On both sides of the Channel by 1700 probably 50 percent of males were in some sense literate. In France the figures were nowhere as high, but after 1700 they were increasing, not stagnating or declining. In Protestant Germany literacy, in the sense of reading, appears to have been common, although by no means a mass or majority phenomenon by the late sixteenth century.[5] The presence of active literacy in early modern Europe is notoriously difficult to calculate, but it would seem to be increasing among urban men and possibly women in the period after 1680. This may have been coupled with a decline or stagnation in the rural or poorer areas of Europe, thereby further widening the gap between elite and popular culture. The crisis of Western culture that gave birth to the widespread assimilation of the new mechanical science was profound precisely because of the presence of a large literate citizenry, greater than any ever assembled since ancient times.

Not least in the course of this crisis stands the failure of the older, scholastic culture to deal effectively with the challenges presented by new experience and empirical data. Indeed by the 1680s it could be said that elite culture had been badly served by the guardians of religious orthodoxy. In the sixteenth and seventeenth centuries both Catholic and Protestant theologians had based the metaphysics of such doctrines as transubstantiation,* consubstantiation, and the Trinity on scholasticism, that is, on the Aristotelianism of the schools. Yet as early as the 1630s, after Galileo's confrontation with the church and the publication of Descartes's *Discourse on Method* (1637), it was clear that Aristotle and Ptolemy no longer adequately described the operations of the natural world, either celestial or terrestrial.

The threat to the metaphysics of orthodox Christianity was real and immediate, yet despite the efforts of Gassendi or the Christian Cartesians to salvage basic doctrines, the clergy of the schools resolutely clung to the old scholastic explanations. By the 1680s one could still find Aristotelians in powerful positions in any school in almost every country, but in western and northern Europe they were on the defensive. They now desperately—

*See glossary of terms.

sometimes mindlessly—sought to maintain doctrinal orthodoxy in the face of the destruction of Aristotelian natural philosophy.

LIBERAL CHRISTIANITY

Eventually the more ingenious clergy, largely of Protestant Europe, realized that it would be necessary to construct a new Christian religiosity based in large measure on mechanical assumptions. That was precisely the synthesis developed by moderate Anglicans, who had been forced under the impact of the English Revolution to rethink the relationship between the natural order, society, and religion. Eventually all progressive European Christians, from the German philosopher Leibniz to the Cartesian priest Malebranche, would be forced to restructure the philosophical foundations of Christianity to conform to one or another version of the new science. It is hardly surprising that liberal Anglicanism, wedded as it was by the 1690s to Newtonian science, took the lead in this enterprise. The Boyle lectures were quickly translated into a variety of Continental languages, and Samuel Clarke became the leading theologian of the godly version of the Enlightenment. Later in the eighteenth century Rousseau invoked the teachings of Clarke, whereas French materialists such as Baron d'Holbach saw him as one of their prime enemies.

In tandem with this liberal version of Christianity came the science of Boyle and Newton. In stark contrast to the doctrinal rigidity of the French Catholic church, or of fundamentalist Calvinism, the English theologians in the tradition of the Cambridge Platonists preached a natural religion founded on reasonable expectations of salvation in the afterlife and reward in this life. The laws of science vindicated God's existence, and the instilling of belief in order, both social and natural, took precedence over such doctrines as transubstantiation and the Trinity. Suddenly a version of Christianity emerged that focused on achievements in this world, on a Christianized self-interest; and this version also embraced the physical universe delineated by the new science.

After 1689 this liberal Christianity became associated in the minds of Europeans with two extraordinary developments. The first was a successful and bloodless revolution in 1688–1689 that removed an absolutist king, James II, established parliamentary sovereignty, and forced the Dutch stadholder, William of Orange, to accept a Bill of Rights as one of the conditions for his ascent to the English throne. It also established a limited religious toleration for all English Protestants, although not for Catholics or anti-Trinitarians. The second innovation was Newtonian science. In the 1690s liberal Anglican clergymen championed both the political settlement of 1689 and the Newtonian synthesis, and related one to the other.

Suddenly the new consensus forged in England stood in stark contrast to the capriciousness of Continental absolutism. A viable national and Anglican church remained amid limited religious toleration, clergymen offered justification for revolution and constitutional government, and a new experimental science had uncovered previously unknown and universal laws. The Newtonian system of the world could be championed as the model for the stable, harmonious, moderately Christian polity ruled by law, not by an arbitrary and capricious will. This polity was the creation of the parliamentary class: large landowners, prosperous merchants, voting freeholders. Political revolution against absolutism had been achieved without social upheaval, without an uprising of the lower orders. Not least, the alliance of England and The Netherlands against the French colossus proved effective. By 1710 the French king had been humiliated on the battlefield; his treasury also stood empty. In this period we should never underestimate the cultural implications of military victory or defeat.

THE POWER OF SCIENCE

Yet before we explore this triumphant Newtonian Enlightenment, as well as the radical alternatives offered to its theism by materialists, pantheists, and atheists—who were themselves enamored of the new science—we should examine the variety of uses to which scientific knowledge was put during the crisis of the late seventeenth century. The psychological epicenter of that crisis lay predictably within Protestant culture, where the traditional responses of piety, prayer, and biblical prophecy came to be seen as increasingly inadequate.

We may take as typical of that older Protestant tradition the mentality of a Protestant Dissenter* of southern England, one Samuel Jeake (b. 1652). He was a highly literate merchant who read widely, whose family favored the Puritan side during the civil wars and who, himself, supported the revolution of 1688–1689. From an early age he began to interpret the events in his life, as well as revolutions in the polity, in astrological terms. This is not to say that he knew nothing of the new science; he certainly read William Harvey's treatise on anatomy and the circulation of the blood. But the culture of the Royal Society or the publication of Newton's *Principia* (1687) passed him by completely. In the early 1690s he found himself on the defensive as he tried to justify astrology "experimentally" and to show that events in 1688–1689 conformed to the radical alterations of the planets in those years.[6] Yet his faith in astrology never wavered; nor did his belief that the events in his lifetime had somehow been foretold by the scriptural prophecies.

In the 1690s such responses were still commonplace. The Huguenot minister Pierre Jurieu and his followers identified Louis XIV as the Antichrist of biblical prophecy and predicted his demise. Jurieu cast a cold eye on natural religion or indeed on any version of Protestant rationalism that denied the clergy an independent and dominant role in the state. He persecuted the Huguenot refugee journalist Pierre Bayle, who in turn used his encyclopedic *Dictionnaire historique et critique* (1697) to mock the doctrinal rigidity of those who predict the future, as well as to scorn absolutism.[7]

Laymen like Bayle, so typical of the crisis, embraced the new science—in his case in its Cartesian form—as an antidote to the scholastic pretensions of the orthodox clergy, both Catholic and Calvinist. Of course this new encyclopedic mind with its passion to order and classify was deeply indebted to the Baconian method of classification and collection. Bayle was the foremost encyclopedist of his age, and where we find his *Dictionnaire* being reedited or imitated in the eighteenth century we will also find journalists with a keen interest in the new science.

The prophetic vision of Francis Bacon, his call to classify all knowledge, did not require the science of Boyle or Newton for

*See glossary of terms.

its survival or application. It appealed to the organizers of knowledge, those directly connected with the printed word, who by the late seventeenth century faced the monumental task of simply keeping track of all that was now being published. The line of influence from Bacon to the great encyclopedia of the Enlightenment, Diderot's *Encyclopédie* (1751), lies through the world of journalists and publishers, particularly in France and the Dutch Republic, who were forced to devise cataloguing and classificatory systems just to keep abreast of their inventory.[8] Their receptivity to the new science lay partly in its salability, but it also grew out of their own sense of the necessity to order the world around them. The inordinate number of Huguenot refugees—many of them once Parisian booksellers—drawn after 1685 to the freer presses of England and The Netherlands meant that presses receptive to science were also the centers for antiabsolutist and anti-Catholic propaganda.[9]

Among the most important Continental promoters of the new science, particularly in its English form, was the liberal Arminian* minister and journalist Jean Le Clerc. His *Bibliothèque universelle et historique,* published in Holland, disseminated news of the *Principia* to thousands of French readers, and it also championed the liberal Christianity of the Anglican moderates. In addition Le Clerc embraced the epistemology of his friend John Locke, whose *Essay Concerning Human Understanding* (1690) had proclaimed the senses as the starting point of all knowledge.[10] Locke had come to intellectual maturity deeply under the influence of the new science as explicated by Robert Boyle, and not surprisingly his philosophy laid emphasis on the external, physical world as the starting point of all knowledge.

Locke and his place within international Protestant circles are perfectly symbolic of the European crisis and its resolution. In the 1680s as an English opponent of royal absolutism he fled for safety to the Dutch Republic. There he became closely associated with liberal Calvinists such as Le Clerc as well as with English refugees of radical background, such as the Quaker merchant Benjamin Furly. Together they discussed every aspect of the contemporary scene: the threat from France, the danger of invasion, Locke's ideas on parliamentary sovereignty, which he had

*See glossary of terms.

worked out primarily in the early 1680s when James II's ascent to the throne appeared inevitable.[11] Not least, he and Furly were familiar with the latest medical theories and with medical reformers who attempted to apply both mechanical and hermetic theories to their practice. In Locke, who was an accomplished doctor, and his circle we see the confluence of interest in the new science and hostility to doctrinal rigidity and absolutism—in short, the Enlightenment in embryo.

Likewise in that circle during the 1690s we can see the confusion of possibilities open to the educated layman in search of alternatives to rigid orthodoxy and authoritarianism. The influence of hermeticists, such as the alchemist F. M. van Helmont (d. 1698?), was still in evidence;[12] indeed Furly himself believed in the mystical doctrine of metempsychosis—that is, the migration of souls after death. By contrast Le Clerc and liberal Dutch theologians argued for science, theism, and toleration, while young refugee journalists, who would later edit Bayle and become radical materialists, found Furly's circle and his library an interesting place to congregate.[13] All were drawn together by the wars against France (1689–1697 and 1701–1713) and the very real fear of a French invasion. In their midst English radicals of freethinking inclination such as John Toland (see pp. 96–97) visited and sought converts; the radical version of the Enlightenment mingled in its early years with moderacy.

At this stage in its cultural history scientific knowledge was still very much a matter of philosophical principles, cosmologies, and rules of reasoning. It was part of the search for an alternative synthesis among educated laymen, doctors, merchants, journalists, politicians, and liberal clerics, for a way out of the crisis provoked by clerical and monarchical authority. Science had not yet become, as it would very shortly, by the 1720s, a body of learning for laymen to master, practice, and apply. Yet to what extraordinary uses natural philosophy could be put within the context "of an entire philosophical liberty," as one of Furly's circle described the desired effect of the "mighty Light which spreads itself over the world especially in those two free Nations of England and Holland."[14]

An example of that light is the assault on magic and popular superstition—as elite culture defined it—launched by the scientific rationalist and Dutch Calvinist minister Balthasar Bekker.

From his vantage point as a citizen of a vast seafaring empire, Bekker compiled a massive catalogue of the superstitions and magical practices found both at home and abroad. Introduced to rationalism and the new science by reading Descartes, Bekker embraced both while retaining his own version of Christian orthodoxy. Where the Bible speaks in the language of the people, as for example in asserting geocentricity, Bekker simply dismissed its cosmology as rhetoric necessary to keep the attention of the common folk. Likewise he denounced the Catholic doctrine of transubstantiation as simply unreasonable.[15] Where the Bible clearly speaks in the voice of God, as in the prophecies that describe the time and circumstances of the end of the world, it must be taken literally. Fideism and Cartesianism mixed in Bekker's mind—so typical of the transition we are describing—in such a way as to permit him to render a major theoretical assault against magic and still remain a millenarian.[16] His *De Betoverde Wereld* (*The World Bewitched,* 1691) was dedicated to a mathematician and the *burgermeester* (mayor) of his native city, Franeker, and it pitted the mechanical philosophy of Descartes against all *tovery en spokery* (witchery and spookery). Bekker sought "to banish the devil from the world and bind him in hell so that the King Jesus might rule more freely." It also labeled the Catholic church as the kingdom of the devil.[17] When he translated his textbook on magic into French, Bekker muted that outright assault on Catholicism and confined himself to attacking the superstitions of "popish" priests. Bekker's French text took its place among a number of such assaults on popular religiosity emanating from French rationalist as well as Jansenist circles. Once again the new science was enlisted against the pagan naturalism of the people,[18] and Bekker's text became a standard and widely read work of the early Enlightenment. We may see him as a transitional figure, not unlike Newton, who combined scientific rationalism with intense religious piety and millenarianism. Yet both thinkers addressed themselves to the laity or to liberal clerics; indeed as might well be imagined, Bekker quarreled with other Calvinist clergy. We might also recall that Newton limited his anti-Trinitarianism to discussions with those clergy who were his Newtonian followers and, possibly, John Locke.

In France Cartesianism had been used by clerical supporters of absolutism to render glory to the Sun King[19] (see pp. 64–66).

Indeed, a French Huguenot was one of the first to attack those scientific mandarins and to argue that science should serve other, more humane ends.[20] But in the hands of Dutch Protestants such as Bekker we can see where Cartesian science might have led had it not been for its negative associations with French absolutism and for the lingering fears about the materialistic implications of Cartesian matter theory.

Indeed, throughout the seventeenth century doubts lingered among the clergy about the meaning of Descartes's radical separation of mind from body. Dutch anti-Cartesians had been vociferous in the 1640s about the danger of materialism (see pp. 68–69). The Cambridge Platonists of the 1660s had sought for similar reasons to lead a new generation of students away from the French philosophy (see pp. 85–87); and as late as 1671 the Scottish Cartesians in Edinburgh actively taught Descartes while still warning against atheistic attempts to use the mechanical philosophy to undermine religion.[21] All those warnings, however dire, preceded the impact made by the Amsterdam philosopher Benedict de Spinoza (1632–1677).

SPINOZA AND SPINOZISM

Born into a family of recently immigrated Portuguese Jews and the son of a merchant, Spinoza read Descartes as part of his school education. As one contemporary biographer put it, from Descartes Spinoza learned "that nothing ought to be admitted as true, but what has been proved by good and solid reason."[22] Spinoza combined his reading of Descartes with deep knowledge of classical and Hebraic texts. Out of that mélange he forged a solution to the Cartesian separation of mind and body that possessed devastating implications for all forms of organized religion. The fears of a century were actualized in Spinoza. He constructed a natural, philosophically anchored version of the human and natural worlds that John Toland, the English radical and follower of Bruno, labeled pantheism.[23] Briefly stated, Spinoza asserted the existence of one infinite substance in the universe, namely Nature or God. He argued that it is illogical or contradictory to posit two kinds of substance, as did all traditional Christian metaphysics, in other words to posit the infinity of God and the separate finiteness of matter. In true Cartesian fashion Spinoza pursued his reason to its clear and distinct conclusion. In the *Tractatus Theologico-Politicus* (1670) he presented his pantheism in an eminently readable fashion and linked it with a philosophy of total freedom from intellectual constraint and with his republicanism.

In the midst of the crisis of the late seventeenth century, spinozism proved the most virulent heresy, and its debt to the new science was inescapable. Spinoza accepted all the Cartesian and mechanical definitions of matter and motion. He then perversely collapsed matter into spirit, God into nature, and a nightmare for Christian natural philosophers became a reality. First Hobbes, then Spinoza—very different, to be sure, in their philosophies of government—and both were so comfortable with purely naturalistic, materialistic, pantheistic explanations of man, society, and nature. Whatever adjective we use should not obscure the adjective most commonly used by contemporaries: atheistic.

To this day the spinozism of those decades possesses a murky history. To its opponents it was everywhere; yet just try to find an avowed spinozist. The Dutch Republic spawned the heresy, and there it can be found among very private circles of professional men, not least of them publishers and journalists. They invented and circulated clandestine treatises that proclaimed Jesus, Moses, and Mohammed as imposters; they championed all science and taught themselves mathematics. They were republicans and critics of monarchical authority; and they had little use for the clergy of the Dutch Reformed Church, who had the right to interrogate heretics and force the authorities to do something to silence them. In their private letters spinozists described the providential God of Christianity as "the god of the lazy."[24] In short they took care of themselves in a competitive world and never resorted to traditional piety for solace. We know of a postal official and servant of the Austrian administration in Brussels in the early eighteenth century who was a spinozist. It all sounds harmless enough from this distance until we realize that this official almost certainly assisted his publisher friends to ship clandestine and heretical literature into France, to undermine the authority of both its church and state. Eventually, in the 1740s, he lost his job because he could not resist publishing and circulating himself yet another heretical piece, a Jansenist work that attacked absolutism.[25]

Jobs and careers could be lost in this period if an individual was accused of irreligion, especially of spinozism. In 1668 an Amsterdam lawyer and doctor who belonged to Spinoza's circle and publicly blasphemed the Judeo-Christian tradition got a sentence of ten years in prison; he died there the following year. In the early eighteenth century Tyssot de Patot, a professor of natural philosophy and mathematics at Deventer in The Netherlands, lost his position for holding heretical views and was ostracized from polite society. Not incidentally he had known Toland in The Hague, where they exchanged clandestine manuscripts—a form of communication for heretical thought that became commonplace during the Enlightenment.[26] An English deist, Thomas Woolston (d. 1733), who challenged the authority of the Bible in matters miraculous and prophetic, died in prison. In Paris during the 1720s the authorities closed down an aristocratic club, L'Entresol, because its members toyed with spinozism and free thought. The leading Continental Newtonian of the first half of the century, William Jacob s'Gravesande (see pp. 185–187), was accused of Spinozism—a heresy to which he did not subscribe but of which he might be accused by devout Dutch Calvinists simply because of his intense involvement with the new science. His successor to the chair of natural philosophy at the University of Leiden, J. N. S. Allamand, was similarly accused although he too successfully protested his innocence.[27] In Leipzig, one of the cultural centers of Protestant Germany, official censors persecuted publishers and booksellers with special zeal when they were suspected of distributing spinozist literature, but also for selling that antimagical text of Bekker. The orthodox clergy regarded any attack on the power of spirits to be tantamount to undermining all spirituality.[28]

The implications of spinozism threatened secular as well as clerical authority, and not simply in the absolutist monarchies. The specter of leveling—as remembered from the English Revolution—lurked about in the early eighteenth-century circles of London deists and freethinkers, many of whom adopted the naturalism of either Hobbes or Spinoza. An anonymous freethinking poem of the 1730s wittily summed up that aspect of the spinozist legacy: Add "mind" (or spirit) to "Nature" and "this Mighty Mind shall be / A Democratic Deity/ . . . all of which we behold is God,/From Sun and Moon, to Flea and Louse/And henceforth equal—Man and Mouse."[29] Freethinkers of this period could hold to ideas with democratic implications while still having little use for the people or their clergy.

THE NEW CULTURE OF THE LAY ELITE

The crisis of the late seventeenth century brought to a head the longstanding tension between the new learning, particularly the new science, of the educated laity and the doctrinal rigor of the traditional clergy. By and large the latter lost the struggle. They could no longer control the printing presses, particularly in England and the Dutch Republic; nor could they eradicate the demand for books and learning, the ever-expanding market for knowledge. The crisis also starkly exposed the dangers to traditional religiosity to be extracted from the new science. Indeed, as a result of the crisis a new cultural persona emerged, first in England and then in western Europe: the literate gentleman who read the periodical press, attended literary and philosophical lectures or clubs for the purpose of being cultured, and remained vaguely Christian, generally Protestant, but explained his beliefs in terms of the order and harmony of creation. He might be a merchant of the city or a landed gentleman of the country; he might even be a shopkeeper, a doctor, or a lawyer. He believed in educating his children; his wife was almost certainly literate and a reader of books, especially novels.[30] By the 1720s, particularly in England, such a gentleman could have increasingly easy access to applied science as taught by the Newtonian lecturers. By the 1760s his son might be investing in industrial adventures or possibly even be an industrial entrepreneur himself. Liberal Protestantism and science made it possible for such men to explain the world to themselves and to feel comfortable in it; eventually, applied mechanical science also made it possible for them to exploit it.[31]

Occasionally one of these literate gentlemen might slip over the edge into outright atheism, generally into pantheism or materialism. We can be sure that when such a conversion occurred it was made easier by familiarity with the principles of the new science. It may even have been prompted by taking Descartes's method of reasoning too literally or by assuming, as did Toland, that Newtonian gravity was a sufficient explanation for the workings of the universe and that no deity other than Nature was necessary. In England such a radical departure from the prevailing cultural wisdom was often accompanied by opposition to the ruling oligarchy or to any version of the old order as it manifested itself at home or abroad. Where we find such radical groups in the late eighteenth century they will be frequently at the forefront of industrialization (see pp. 164–168). For them science was tied to a larger vision of social reform through the application of machinery to production. Such radical gentlemen embraced capitalism perhaps even more willingly than did their more moderate counterparts. For those who could control it, capitalism of an industrial and mechanical sort represented, on both sides of the Channel, an effective means of destroying the monopolies exercised by the old landed aristocracy.

Whatever one's private religiosity or politics became, by the early eighteenth century it bore little resemblance to the expressive piety of popular Catholicism or to the rigorous demeanor of orthodox Calvinism. Any kind of sectarian "enthusiasm," the public preaching of millenarians who announced the end of the world or the ecstasies of parishioners who thought they had discovered a saint in their midst—such an event occurred in Paris in the 1720s—was an object of scorn and derision on the part of enlightened men, and also quite possibly of enlightened women.[32] The scorn for "the inferiour herd of people" was endemic to enlightened culture; "our people of the lowest rank, for want of due care to instruct them, are worse than Hottentots," as one freethinking English journal put it. The only remedy was to instill "the most familiar and evident truths in natural philosophy . . . some of the fundamental maxims of a free-government . . . and practical precepts of religion and morality." These alone might "dispose the people to virtue, without which we can never long continue a flourishing nation."[33]

Throughout western Europe educated elites eagerly absorbed the scientific learning coming from England, the science of Newton and the Royal Society. In 1700 Pierre Bayle urged a bright young Huguenot refugee with an interest in science to go to England: "It is the one country in the world where profound metaphysical and physical reasoning is held in the highest regard."[34] By that year Newtonian science had also attracted followers in the Dutch Republic, especially at the University of Leiden (see pp. 185–186) as well as at the French language presses run by Huguenot refugees or Dutch Arminians. It offered a new yet moderate synthesis that eschewed materialism but implied a tolerant and progressive way out of the crisis in confidence that had afflicted elite culture. The fact that these receptive groups were already present, and decidedly not from the mainstream of traditional Dutch Calvinism, indicates the irreconcilable divisions within European Christendom that the crisis now finally exposed. Nor is it accidental that non-Calvinist Protestants who emphasized the right of the individual to find his or her own salvation, for example, the Mennonites, were particularly receptive to the new science throughout the late seventeenth and eighteenth centuries. The leading liberal Mennonite theologian of the Dutch Enlightenment, Johannes Stinstra, adorned his wall with a portrait of the Newtonian philosopher Samuel Clarke, whom he had translated.[35]

But it was the French language press of the republic, edited by the Dutch and Huguenot journalists, that first announced Newtonian science on the Continent. The pages of the *Journal littéraire* (1713–1732), the *Nouvelles de la république des lettres* (1700–1710), *L'Histoire critique de la république des lettres* (1712–1718), *Nouvelles littéraires* (1715–1720), and the *Bibliothèque raisonnée* (1728–1752) are bursting with English culture, but particularly with explications of liberal Anglicanism and the latest scientific publications. In addition the Dutch theologian Bernard Nieuwentyt wrote one of the most important textbooks on this liberal and Newtonian theology, *The Religious Philosopher* (1715), which after its translation became a standard text in English schools; it was also popular in French and German translations. It was unrelenting in its attack on spinozism, and it presented a mélange of science and religion—known at the time as physico-theology—that emphasized the harmonious and hierarchical order of nature and society. Significantly the English translation done under Newtonian auspices removed Nieuwentyt's extensive references to the Bible.[36] A century of doctrinal quibbling had convinced liberal Protestants that science was a better anchor for religion than either of the Testaments.

In England the first generation of Newtonians—Richard Bentley, Samuel Clarke, John Derham, and William Whiston—had taken Newton's science into the pulpit. Yet as early as the 1690s Newtonian science, or more precisely the new mechanical science as synthesized by the *Principia,* was also unveiled in far more secular settings. In coffeehouses and printers' shops, Newtonian explicators such as John Harris, Francis Hauksbee, and Whiston assembled audiences and gave "a course of Philosophical Lectures on Mechanics, Hydrostatics, Pneumatics [and] Opticks."[37] Such lectures frequently received aristocratic patronage and became very much a part of the culture of the ruling Whig* oligarchy.

Indeed the linkage between the promotion of Newtonian science and the interests of that oligarchy was by no means accidental. The latitudinarian hierarchy of the established church, much to the horror of the lower clergy, gave its blessing to the triumphant Whig party. The scientific ideology of order and harmony preached from the pulpits complemented the political stability over which that oligarchy presided. At the Royal Society the followers of Newton, partly as a result of his direct influence, were firmly in control of that institution and kept anti-government or Tory dissidents out of positions of authority. By the 1720s and the accomplishment of the Hanoverian Succession (1714)—which ensured the survival of Protestant monarchy, the Whig party, and the established church—a new generation of Newtonians had come to prominence and very much set the terms of the Enlightenment in England.

At the Royal Society under the leadership of such Whigs as Martin Folkes and Sir Hans Sloane scientific application to industry and commerce—always a part of its mission—took on increasing prominence. Likewise we see an easing of the doctrinal preoccupations of the first generation of Newtonian clergymen; indeed Folkes and his friends appear to have had little interest in organized religion.[38] Newtonianism supplied all the answers they needed in order to live lives of relative comfort amid the prosperity and political stability enjoyed by the upper classes—

*See glossary of terms.

and some middling folk—in the Hanoverian* state. The gardens of Queen Caroline at Richmond contained busts of Newton, Locke, Clarke, Boyle, and the liberal theologian William Wollaston, which expressed her faith in Newtonian science and natural religion.[39]

Typical of this Newtonian culture, with its emphasis on practical science, is a text like Henry Pemberton's *A View of Sir Isaac Newton's Philosophy* (1728). It is a much more straightforward and succinct account of Newton's philosophy of nature, his definitions of matter, space, time, the vacuum, and the law of universal gravitation, than that found in the Boyle lectures. Christian apologetics have been de-emphasized in favor of a general, but constant, emphasis on the power of the deity, on a straightforward explanation of Newtonian physics. Whenever Pemberton enters into polemics, it is only against the materialists: those who assert that gravity is essential to matter, those who would have the immortality of the world, those who deny the supremacy of God in every aspect of creation. This fashionable Newtonian and providential "deism" had now replaced the doctrinal exactness of the early Newtonians.

By far the most famous transmitter of this Newtonian culture to the Continent was the French poet and philosopher Voltaire. When he arrived in London in 1726 he learned Newtonianism directly from Samuel Clarke, and for Voltaire it took on the force of a religion.[40] His *Lettres philosophiques* (1733), an immensely popular paean of praise to English government, mores, and science, linked the achievements of Newton to a milieu of intellectual liberty such as existed, he claimed, only in England. He offered English science and society as a universal model for enlightenment, and in the process he further secularized Newtonianism. He insisted on the existence of Newton's God; but in Voltaire's hands the concept becomes largely impersonal, its function could be described as simply social. The deity maintains order and so too should monarchs and governments. The English aristocracy is praised precisely because of its willingness to be educated, to mix with men of learning and science. Like the English Newtonians, Voltaire repudiated the science of Des-

*See glossary of terms.

cartes, and for similar reasons. Not only is it wrong, but his private notes on the matter tell us that it leads directly to materialism and atheism.[41] Voltaire made Newton and his science fashionable, and he linked both to his rabid anticlericalism and his denunciation of superstition and intolerance. The new science, he proclaimed, is the alternative to priestcraft and bigotry. The argument became famous by the 1740s.

Beginning in the 1690s English and then Continental Newtonians undertook a vast propaganda campaign against Cartesian science. For someone like the Dutch doctor and professor Boerhaave, the science of Descartes was insufficiently experimental, for others the primary fear was that Cartesianism led directly to materialism. Voltaire put his objections succinctly:

> With regard to the pretended infinity of matter [for Descartes matter is extension], that idea hath as little foundation as the vortices. . . . But what are we to understand by an infinite matter? For the term *indefinite,* used by Descartes, either must be explained by this, or it signifies nothing at all. Do they mean, that matter is essentially infinite in its own nature? If so then Matter is God.[42]

Voltaire's deism rested on the assumption that "God the General in the Universe gives orders to different bodies."[43] Without those orders there can be no order. Voltaire believed that without God nothing would restrain kings or impose order on the masses. Any explanation for the triumph of Newtonian science in the early eighteenth century that ignores or underestimates the force of these social and ideological considerations misses the cultural context within which science, like any other body of knowledge, had to be mediated.

Perhaps of even greater importance than Voltaire as a transmitter of Newtonian science within western Europe was the Dutch scientist s'Gravesande (1688–1742) (see pp. 185–186). His *Mathematical Elements of Natural Philosophy* (Latin edition, 1720–1721, English, 1720–1721 and five subsequent editions; French, 1746–1747) gave a sophisticated explication of Newtonian science in textbook form that was never surpassed in the first half of the century. In 1717 at his inaugural lecture for the professorship in astronomy at the University of Leiden, a post secured for him through Newton's intervention, s'Gravesande defended mathematicians from the accusation of atheism and irreligion. He also lashed out at "those men who have never thought that their very existence and that of the things around them would not be possible without the effects of a powerful and a very wise Cause . . . and against those who are only occupied with religion as it is an object of their indecent railleries." s'Gravesande always maintained the Newtonian objection to materialism; and in general his text, although more mathematically sophisticated, is similar to Pemberton's. He eschewed the polemics of the first generation of clerical Newtonians and concentrated his attention on explicating the *Principia*. Visitors to his university in the 1720s saw proudly displayed in its library "a fine brass sphere which shows the motion of all the planets according to the Copernican system and is moved by a pendulum."[44] Indeed with devices and textbooks such as these, especially when combined with the easier texts in Newtonian science that became increasingly commonplace (see Chapter 5), the *Principia* could be safely ignored by those in search of a basic scientific education.

The Birth of European Freemasonry

It is little wonder that this cultural synthesis based on science, religion, and social ideology, which was preached from the fashionable London pulpits and published in fancy editions partly financed by lawyers, merchants, and Whig members of parliament,[45] should also produce a new form of social gathering complete with ritual and costume. British Freemasonry began in 1717 as a speculative, gentlemanly club quite different from the older masonic guilds from which it originated. The practicing masons along with their culture had been totally displaced—indeed the concept of the guild protecting the wages of its workers is self-consciously repudiated by the new masonic *Constitutions* (1723). In their place have come the scientific devotees with one out of four Freemasons in the 1720s being fellows of the Royal Society.[46] Indeed one of the most active Freemasons in its early years was the Newtonian scientist and experimenter Jean T. Desaguliers (see pp. 143–144). He was undoubtedly instrumental in spreading the society from London to the English provinces and the Low Countries.[47]

At the masonic gathering, that quintessential popularization of enlightened culture, literate gentlemen of substantial means (one had to afford the dues) worshiped the "great Architect," the god of the new science, and gave allegiance to any religion they cared to name: "to the religion of that Country or Nation, whatever it was, yet 'tis now thought more expedient only to oblige them to that Religion in which all men agree, leaving their particular opinions to themselves."[48] Armed with the principles of geometry as well as "the Mechanical Arts," "several noblemen and gentlemen of the best rank, with clergymen and learned scholars" constituted lodges where "all preferment" is based on "personal merit only." In some of the earliest British lodges scientific experiments were performed and lectures on the new science given.[49]

The lodges as they spread to both sides of the Channel should not be understood, however, as centers of scientific learning. They were primarily social clubs that gave ritualistic expression to the fraternity of the meritorious and encouraged them to improve their literacy, education, and decorum. The lodges sometimes kept libraries or sponsored reading societies; not accidentally, Freemasons in eighteenth-century Europe were active in promoting scientific education in excess of their numbers. When Desaguliers lectured on mechanics in Rotterdam, Amsterdam, The Hague, and Paris—speaking in English, Latin, or French—he undoubtedly attracted men who in turn sought out membership in his fraternity (women were permitted to attend these scientific lectures and did, but were excluded from Freemasonry).

Brothers in this fraternity imbibed a mythic lore that connected their society back to ancient times and thereby stamped as ancient, and hence respectable, that which was irredeemably modern and, in the first instance, British. The cultural persona given expression in the British lodges was overwhelmingly bourgeois: lawyers, merchants, civil servants, publishers, and gentlemen of science abounded. On the Continent aristocratic leadership became widespread, and the lodges tended to mirror the social stratification endemic throughout the old order. Yet the British roots of Freemasonry could not be totally obscured.

Such an essentially Protestant archetype did not necessarily travel well on the Continent. After masonic meetings were held in the 1730s in the home of the British ambassador in Paris, the police raided his headquarters and insisted that the meetings stop. When Freemasons journeyed to Portugal and Italy they were arrested and even tortured by the Inquisition. It sought to discover their "secrets," which amounted to nothing more than old rituals inherited from the guilds and a rather flexible definition of what constitutes religion. While in time masonic lodges under aristocratic patronage flourished in Catholic Europe, the fraternity as a whole was condemned by the pope in 1738.[50]

The papacy, not without cause, spied the makings of a new religion. For dissidents of church and chapel, for opponents of established authority, the masonic lodges offered an alternative society where any heresy might be freely discussed. The leading Amsterdam Freemason of the 1730s and 1740s was, not accidentally, a self-proclaimed pantheist who adored the new science and believed that "Nature . . . places us willy-nilly on this earth, not forever but for a limited time, whose extent and final end are alike hidden from us; this is the universal order to which everyone, but especially men of reason, do well to submit themselves."[51]

This extraordinary faith in the order and reasonableness of nature, as proclaimed and mediated by science, might also make radicals out of those who took it too seriously. Eighteenth-century society and government, especially on the Continent, was at best profoundly oligarchical and at worst rigidly hierarchical and totally nonrepresentative of mercantile or industrial interests and values. Inevitably there are links between the synthesis of science and religion that resolved the crisis at the beginning of the eighteenth century and the revolutions coming later, first in the American colonies (1776), then in Amsterdam and Brussels (1787), and finally in Paris (1789). A progressive faith born out of the new science and sustained by its achievements rendered enlightened men potentially impatient, even rebellious, in the face of practices or old elites disinterested in improvement and economic growth based on the freedom to trade, or worship, or experiment. Late in the eighteenth century some such men remolded the masonic lodges into places where their radicalism and impatience could gain expression. The new lodges probably had little resemblance to those founded earlier in the century by Desaguliers. Yet in ideological terms they remind us that the belief in progress promised by the new science might lead to expectations never intended by its original exponents.

The Application of Newtonian Science

Newtonian science in the hands of the laity was, however, more than ideology. It was also, and increasingly, practice. Prior to the assimilation of the *Principia,* mechanics certainly existed as a body of science capable of application; but what was lacking was any overriding theory or set of principles, a natural philosophy and set of laws to give it coherence. We can compare a pre-Newtonian textbook of good practical mechanics with what came immediately after it. Such manuals were frequently anti-Aristotelian but could offer no coherent alternative explanation of gravity, although they were perfectly adequate in explaining how levers, wedges, and pulleys could be employed.[52] As one historian of science has put it, "the parallelogram of forces, the law of the lever, the principle of virtual work, the action of contact forces, and the principle of energy had extensive earlier histories," but all of these aspects of classical mechanics "were to be absorbed in or united to the Newtonian stream."[53] From the more general perspective of Western culture, that body of mechanical learning also received an unprecedented public exposition after the publication of the *Principia* (1687).

In the first Newtonian lectures ever given, Francis Hauksbee explicated "the general laws of Attraction and Repulsion, common to all matter," as proclaimed in the *Principia,* which thereby "establish . . . the true system of Nature, and explain . . . the great motions of the world." Then follows a detailed description of Boyle's air pump as a machine "for giving a swift motion to bodies in vacuo." Hauksbee possessed a particular interest in the phenomenon of "action at a distance," of which electricity was the most fascinating and spectacular example. Those attractive electrical forces are defined as essentially an aspect of the overall "power in nature by which the parts of matter do tend to each other"—in short, another illustration of Newton's principles. Throughout these lectures mechanical devices are used to illustrate the laws of Newtonian science, and the emphasis is on perfecting those devices.

In Hauksbee's lectures no direct industrial application of those machines is made, although tables are given for the specific gravity of stone and coal found commonly in the mines of the Midlands.[54] The coal mining of Britain was by 1700 the most advanced in Europe. The output of coal in France at the end of the seventeenth century probably did not amount to more than 75,000 tons a year, which was less than had been mined on a single north country manor prior to the English Revolution.[55] On the Continent only Belgian coal production came close to the English figures, and predictably both s'Gravesande and Desaguliers were active in Belgium (then called the Austrian Netherlands) in the 1720s attempting to install steam engines to drain those mines.

The application of the new science could hardly be resisted; indeed it had been encouraged by the scientists of the Royal Society as early as the 1680s (see pp. 92–93). In that period such applications were more desired than possible; but the commitment to render science useful to trade and industry became a part of the ideology of English science from the 1660s, if not earlier. After 1700 ideology came to bear relation to reality, and simultaneously the lecturers of the London coffeehouses moved into the provinces: to the north, Newcastle-upon-Tyne in 1711–1712, Derby in 1728, to the midlands, Peterborough and Stamford, in the 1730s. The provincial academies, the schools of the non-Anglican Dissenters, also eagerly took up that science and replicated those lectures in their classrooms. By 1730, not incidentally, there were over 100 steam engines at work in Britain.

As we shall see, those engines cannot be severed from the diffusion of the English Enlightenment, from the science that lay at the heart of that cultural transformation. It could foster industry just as easily as it could instill a rather cerebral piety. It could edify and instruct the genteel; it could also appeal to provincial entrepreneurs more interested in capital profits than cultural sophistication. Such men possessed a sense of what was happening in the world around them and why it was necessary to educate oneself in science. They bought scientific books and attended scientific lectures in increasingly large numbers. The ease with which Newtonian science could now be taught made irrelevant the ideological struggles and metaphysical disputes that had once dominated natural philosophical discourse in seventeenth-century Europe.

Late in the eighteenth century, at the height of the British Industrial Revolution, mechanical science and the ideology of progress it promoted seemed to the leaders of that economic transformation the answer to all human misery. It would secure their wealth and power eternally while eliminating the excesses of poverty still commonplace to the majority of men and women. Industrialists presumed that "the application of steam to the various purposes contemplated [will] not [be] very difficult," that there will be new machines "with greater the velocity and less the expense." They proclaimed their faith (as a leading industrialist put it)

> in the stupendous effect which the application of mechanical science is on the eve of bursting upon the world—when, in the transport of ourselves—as well as the enormous masses of other hands [i.e., the workers]—time, distance, and expense shall be almost annihilated. This will be laughed at now, as was Sir Richard Arkwright a half century ago, when he predicted Cotton Yarn and Cloth would be sent from here to the East Indies.[56]

The new industrialists gloried in mechanical science; they sent their children only to those universities (Edinburgh, for example) and Dissenting academies where they were sure the most up-to-date state of the art would be taught.[57] Armed with that science they believed it possible "to ameliorate the condition of the great mass of the people not in Europe only but in the World—the rising generation are soon to form that Mass, some will rule and some will obey, but all will in one way or other have influence in the management of affairs."[58] Armed with a science significantly divorced by the early eighteenth century from the culture of the people and their immediate necessities, the first industrialists (not unlike their modern successors) believed that somehow they could retain a social order that primarily rewarded and enriched themselves while still improving the human condition. The dream goes back to Francis Bacon. Its widespread acceptance among the educated elite began only in the early eighteenth century, and so too did modern scientific culture.

NOTES

1. For a brilliant discussion of the crisis, see Paul Hazard, *The European Mind* (New Haven: Yale University Press, 1953).
2. For an analysis of the working of that censorship in France, see

Joseph Klaits, *Printed Propaganda Under Louis XIV: Absolute Monarchy and Public Opinion* (Princeton: Princeton University Press, 1976).

3. For another approach to the emergence of probability, see Barbara Shapiro, *Probability and Certainty in Seventeenth Century England* (Princeton: Princeton University Press, 1983).

4. A good example of the virulence of that campaign can be found in Aubrey Rosenberg, *Nicholas Gueudeville and His Work (1652–172?)* (The Hague and Boston: Nijhoff, 1982), p. 61; Pierre J. W. van Malssen, *Louis XIV d'après les pamphlets repandus en Hollande* (Amsterdam: H. Paris, 1936); Guy Howard Dodge, *The Political Theory of the Huguenots of the Dispersion* (New York: Columbia University Press, 1947); K. Malettke, *Opposition und Konspiration unter Louis XIV* (Göttingen: Vandenhoesch und Ruprecht, 1976).

5. See David Cressy, "Levels of Illiteracy in England, 1530–1730," in Harvey L. Graff, ed., *Literacy and Social Development in the West: A Reader* (Cambridge: Cambridge University Press, 1981), pp. 123–124. On Germany, see Gerald Strauss, *Luther's House of Learning: Indoctrination of the Young in the German Reformation* (Baltimore: Johns Hopkins University Press, 1978), p. 202.

6. Clark Library, Los Angeles, MS J43M3 A859, "Astrological Experiments Exemplified by Samuel Jeake"; cf. his diary, MS J43M3 D540, 1G94.

7. Pierre Retat, *Le Dictionnaire de Bayle et la lutte philosophique au XVIIIe siecle* (Paris: Presse de Université de Lyon, 1971).

8. C. M. G. Berkevens-Stevelinck, *Prosper Marchand et l'histoire du livre* (Ph.D. diss., University of Amsterdam, 1978), pp. 2–16.

9. See Margaret C. Jacob, *The Radical Enlightenment: Pantheists, Freemasons and Republicans* (London and Boston: Allen and Unwin, 1981), Chapter 7.

10. Cf. G. Bonno, "Lettres inedites de Le Clerc à Locke," *University of California Publications in Modern Philosophy,* vol. 52 (1959).

11. On Furly, see William Hull, *Benjamin Furly and Quakerism in Rotterdam* (Philadelphia: Swarthmore Monographs, 1941); for his library, see *Bibliotheca Furliana* (Rotterdam, 1714).

12. See British Library, MSS. ADD. 4283, fols. 265–266, and Furly's letters to William Penn at the Pennsylvania Historical Society, Philadelphia.

13. M. C. Jacob, *The Radical Enlightenment,* p. 218.

14. Public Record Office, London, MS 30/24/22/6, third earl of Shaftesbury to Jean Le Clerc, March 6, 1706.

15. Balthasar Bekker, *De Philosophia Cartesiana admonitis candida et sincera* (Vesaliae, 1668), pp. 14–18.

16. Balthasar Bekker, *Uitlegginge van den Prophet Daniel* (Amsterdam, 1688). The preface is dated May 14, 1688, and is clearly written under the impact of the outfitting of the Dutch fleet for what many assumed would be a war against France. Cf. K. H. D. Haley, "Sir Johannes Rothe: English Knight and Dutch Fifth Monarchist," in

Donald Pennington and Keith Thomas, eds., *Puritans and Revolutionaries: Essays in Seventeenth-Century History Presented to Christopher Hill* (Oxford: Clarendon Press, 1978), pp. 310–332.

17. Balthasar Bekker, *De Betoverde Weereld* (1691), preface and p. 656.

18. Balthasar Bekker, *Le monde enchanté* (Amsterdam, 1694), vol. 4, pp. 296, 719. On journalistic propaganda in support of Bekker, see J. J. V. M. de Vet, *Pieter Rabus (1660–1702)* (Amsterdam: Holland University Press, 1980). Cf. Jacques Revel, "Forms of Expertise: Intellectuals and 'Popular' Culture in France (1650–1800)," in Steven L. Kaplan, ed., *Understanding Popular Culture: Europe from the Middle Ages to the Nineteenth Century* (Berlin: Mouton, 1984), pp. 255–273.

19. See L. Marsak, "The Idea of Reason in Seventeenth Century France: An Essay in Interpretation," *Journal of World History*, vol. 11 (1968), p. 409.

20. Erica Harth, *Ideology and Culture in Seventeenth Century France* (Ithaca: Cornell University Press, 1983), pp. 290–292, 297, on Denis Vairasse.

21. R. H. Campbell and A. S. Skinner, *The Origins and Nature of the Scottish Enlightenment* (Edinburgh: Donald, 1982), p. 70, in Christine M. Shepherd in Campbell and Skinner, eds., "Newtonianism in Scottish Universities in the Seventeenth Century."

22. John Colerus, *The Life of Benedict de Spinosa, Done out of French* (London, 1706), pp. 3, 7. To be used with some caution, as Colerus is an essentially hostile source.

23. On the career of pantheism as derived from Spinoza and others, see Paul Verniere, *Spinoza et la pensée française avant la revolution*, 2 vols. (Paris: Presses Universitaires de France, 1954).

24. M. C. Jacob, *The Radical Enlightenment*, p. 244.

25. See Margaret C. Jacob, "The Knights of Jubilation: Masonic *and* Libertine," *Quaerendo*, vol. 14 (1984), pp. 63–75.

26. Aubrey Rosenberg, *Tyssot de Patot and His Work, 1655–1738* (The Hague: Nijhoff, 1972); and Rosenberg, "An Unpublished Letter of Tyssot de Patot," *Vereeniging tot Beoefening van Overijsselsch Regt en geschiedenis*, vol. 96 (1981), pp. 71–76. Cf. Alan Gabbey, "Philosophia Cartesiana Triumphata: Henry More (1646–71)," in Thomas M. Lennon et al., eds., *Problems of Cartesianism* (Kingston, Ontario: McGill-Queen's University Press, 1982), p. 246.

27. Koninklijk Huisarchief, The Hague, MS G 16-A29, fol. 14, Allamand to M.M. Rey, 1762.

28. Agatha Kobuch, "Aspekte des aufgeklarten burgerlichen Denkens in Kursachsen in der ersten Halfte des 18. Jh. im Lichte der Bucherzensur," *Jahrbuch für Geschichte* (Berlin, DDR), 1979, pp. 251–294.

29. [Anon.], *War with Priestcraft Or, the Freethinkers' Iliad: A Burlesque Poem* (London, 1732), pp. 36–37.

30. Ruth Perry, *Women, Letters and the Novel* (New York: AMS Press, 1980).

31. For a splendid description of this new culture, see Roy Porter, "Science, Provincial Culture and Public Opinion in Enlightenment England," *British Journal for Eighteenth Century Studies,* vol. 3, no. 1 (1980), pp. 20–46. For a fascinating account of the earliest applications of Newtonian science, see Larry Stewart, "The Selling of Newton: Science and Technology in Early Eighteenth-Century England," *Journal of British Studies,* vol. 25 (1986), pp. 178–192.

32. See Margaret C. Jacob, "Freemasonry, Women, and the Paradox of the Enlightenment," in *Women and the Enlightenment* (New York: Institute for Research in History, 1984); cf. Ran Halevi, *Les Loges Maçonniques dans la France d'ancien régime* (Paris: Colin, 1984), a facile but useful study.

33. *The Freethinker* (London), no. 16 (May 16, 1718), pp. 69–72. Cf. Harry Payne, *The Philosophes and the People* (New Haven: Yale University Press, 1976).

34. *Oeuvres diverses de Pierre Bayle,* 3 vols. in 4 (Hildesheim, 1968), vol. 4, pp. 794–795.

35. J. van der Berg, "Eighteenth century Dutch translations of the works of some British latitudinarian and enlightened theologians," *Nederlands archief voor kerkgeschiedenis,* n. s. vol. 59, no. 2 (1979), pp. 198–206.

36. A. C. de Hoog, "Some Currents of Thought in Dutch Natural Philosophy" (Ph.D. diss., Oxford, 1974), pp. 300–301. Jean T. Desaguliers sponsored this edition, and its translator told Toland that it was aimed against him.

37. *The Englishman,* no. 42 (January 26, 1714), cited in James E. Force, *William Whiston: Honest Newtonian* (Cambridge; Cambridge University Press, 1985), p. 162–163n.

38. J. Force, *Whiston,* pp. 135–136.

39. Judith Colton, "Kent's Hermitage for Queen Caroline at Richmond," *Architecture,* vol. 2 (1974), pp. 181–191. Occasionally Newtonians could be Jacobites; see Andrew Cunningham, "Sydenham vs. Newton: The Edinburgh Fever Dispute of the 1690's . . ." *Medical History,* supplement 11, 1981, pp. 71–79.

40. René Pomeau, *La Religion de Voltaire* (Paris: Nizet, 1956).

41. Voltaire, *Traité de Metaphysique (1734),* ed. H. Temple Patterson (Manchester: Manchester University Press, 1957), pp. 17–19.

42. Voltaire, *The Elements of Sir Isaac Newton's Philosophy,* trans. John Hanna (London, 1738), pp. 182–183.

43. Ibid., p. 236n.

44. For s'Gravesande's statement, see J. N. S. Allamand, ed., *Oeuvres philosophiques et mathematiques de M. W. J. s'Gravesande* (Amsterdam: Marc Michel Rey, 1774), vol. 2, pp. 316–317. The sphere was seen by an English woman tourist in 1726, Clark Library, MS J86Z, n.f. Wednesday, 16 June. According to one account, this "fine Copernican sphere with 1500 wheels, made by Tracy an English Man Living

at Rotterdam which not only shews the different motions of the heavenly bodies but the year, month, day and how and which cost 30,000 guilders or £3000"; Los Angeles: Clark Library, MS Phillips 9356.

45. W. A. Speck, "Politicians, Peers and Publication by Subscription, 1700–50," in Isabel Rivers, ed., *Books and Their Readers in Eighteenth Century England* (Leicester: Leicester University Press, 1982), p. 64.

46. J. R. Clarke, "The Royal Society and the Early Grand Lodge Freemasonry," *Ars Quatuor Coronatorum,* vol. 80 (1967), pp. 110–119.

47. See J. A. van Reijn, "John Theophilus Desaguliers, 1683–1983," *Thoth,* no. 5 (1983), pp. 165–203.

48. *The Constitutions of the Freemasons* (London, 1723), p. 50.

49. See M. C. Jacob, *The Radical Enlightenment,* Chapter 4.

50. For Freemasonry in Catholic Europe, see Jose A. Ferrer Benimelli, *Masoneria, Iglesia e Ilustracion* (Madrid, 1976), vol. 2, p. 202 ff.

51. Quoted in M. C. Jacob, *The Radical Enlightenment,* pp. 243–244. The quotation is by Rousset de Missy.

52. V. Mandey, *Mechanick Powers; or the Majesty of Nature and Art Unvail'd* (London, 1702).

53. E. Truesdell, "Reactions of Late Baroque Mechanics to Success, Conjecture, Error, and Failure in Newton's *Principia,*" in Robert Palter, ed., *The "Annus Mirabelis" of Sir Isaac Newton, 1666–1966* (Cambridge, Mass.: MIT Press, 1970), p. 209.

54. Francis Hauksbee, *Physico-Mechanical Experiments in Various Subjects . . .* (London, 1719).

55. J. U. Nef, *The Rise of the British Coal Industry,* 2 vols. (London, 1966; Cass reprint of 1932 edition), vol. 2, p. 126–128.

56. Fitzwilliam Museum, Cambridge, MS 37-1947, William Strutt to Maria Edgworth, 1823. Similar sentiments are to be found in the Strutt MSS, Derby Local Library, Derbyshire.

57. MS 48-1947, manuscript by Joseph Strutt, "On the relative advantages and disadvantages of the English and Scottish Universities," 1808.

58. Ibid.

CHAPTER 5

The Cultural Origins of
the First Industrial Revolution

I scarcely know without a good deal of recollection whether I am a
Landed Gentleman, an Engineer or a Potter, for indeed I am all
three and many others characters by turns.

Josiah Wedgwood to Thomas Bentley, 1765

Many of my experiments turn out to my wishes, and convince me
more and more, of the extensive capability of our Manufacture for
further improvement. It is at present (comparatively) in a rude,
uncultivated state, and may easily be polished, and brought to much
greater perfection. Such a revolution, I believe, is at hand, and you
must assist in, [and] profit by it.

Josiah Wedgwood to Thomas Bentley, 1766

With the assistance of the natural philosophers and their science,
English entrepreneurs such as Josiah Wedgwood and the Strutts
thought their way to industrialization. That intellectual history
has largely gone unwritten because economic and social history,
and modernization studies, have dominated the study of the first
Industrial Revolution.

Certain assumptions about early industrialization have been
commonplace in the recent literature: The mechanistic economic
concepts of "take-off," which once begun becomes irreversible;
the "state of business expectations" (i.e., the necessity to achieve
profit); the growth of population; access to raw materials and
exploitable labor; and not least, a specific "specialization" in
financial and commercial operations—all have been assumed as
prerequisites to the industrial process.[1] Although none of these
factors can be discounted, none of them explain the obvious. The
presence of raw materials, foreign markets, cheap labor, and

surplus capital will not in themselves make an industrial revolution. The entrepreneur, the historical actor, remains critical to the process as does a *mentalité* favoring "modernization"—the historian of the eighteenth century would prefer the term "improvement"; and without these actors the industrialization of the means of production and the transformation in living patterns of both entrepreneur and worker simply will not occur. The recent history of the various attempts to impose industrialization and modernization on third world countries reveals with stark clarity that technology, capital, and imported expertise will not produce the intertwining of political, social, and intellectual factors that in eighteenth- and nineteenth-century Europe permitted industrialization to occur, first in England and then gradually in much of the western and northern areas of the Continent.

The failure of modernization in various contemporary societies must of necessity affect the academic disciplines most directly concerned with explaining economic development historically or with understanding the social relations of science and technology. The time has come therefore to take another look at the first Industrial Revolution, more precisely at the role of science in that process. A careful reading of the sources left by British eighteenth-century natural philosophers and their followers reveals, almost at first glance, that one assumption which used to be fairly common must now be discarded: that "pure" science had nothing to do with industrialization.[2] In the eighteenth century—and this may be the key to the "curious" nature of scientific inquiry in that period—the distinction between "pure" and "applied" science simply did not exist for the natural philosophers themselves. Eighteenth-century scientists, whether as electrical experimenters, mathematicians, or mechanists, moved with ease between theory and application. They did so because they existed within a particular social and political milieu that favored, indeed encouraged, the articulation of scientific knowledge in the service of the literate elite, both landed and commercial.

The political stability so beneficial to that elite rested on certain ideological assumptions derived from seventeenth-century political theory and from the experience of the revolutions of 1640 and 1688–1689. Out of the first of the great modern revolutions developed a political system that displayed certain distinctive characteristics: strong centralized government bound by a constitutional settlement that favored parliamentary rule and ministerial or "court" power over the interests of the church or the people (the franchise being limited to approximately one-fifth of the male population); religious toleration for all Protestants; and not least, a political culture based on publicly held elections and debates and, even more important, on the relatively free circulation of the printed word. From the founding of the Royal Society in 1662 until well into the 1690s, as we have seen in Chapter 3, the new science articulated by Boyle and Sprat and the Newtonians became tied, both in its conceptual assumptions and in its social goals, to the maintenance of strong central, but not absolutist, government and to the articulation of a liberal, but nonetheless Anglican, hegemony. These links between the ideology of the new science as it developed in the Restoration and beyond and the political order as it emerged after 1689 permit us to describe the natural philosophy inherent in Newton's science as the metaphysical foundations of the Whig constitution.

Given these social foundations, the science found most commonly in eighteenth-century England possessed certain characteristics unique to that culture, although this science was always capable of exportation to and modification on the Continent. In the previous chapter we have explored some of the reasons why the new science achieved such widespread assimilation throughout European culture during the first quarter of the eighteenth century. Now we must seek to recapture the dominant, but by no means the only, uses to which Newtonian science was put in its country of origin, in the first country anywhere in the world to experience industrialization.

THE ENGLISH ENLIGHTENMENT

Having described England in that manner we run the risk of looking far too early in the century for science specifically linked to industrial needs and aspirations. That sort of hindsight may be avoided by approaching eighteenth-century England in the first instance from the perspective of the general historian, and asking of it the questions most perplexing in current historical discussion. Having ushered in, as it were, identifiably modern forms of political organization, religious toleration, and relative intellectual freedom, did England experience the cultural movement that it inspired, and can we speak of a specifically English, as opposed to French or German or Scottish, Enlightenment? And finally, having captured the intellectual leadership of the Scientific Revolution in the second half of the seventeenth century, what happened to English science in the period after Newton?[3] Did it in fact decline as some historians have contended?

The answer to these questions lies in an examination of eighteenth-century English scientific culture. There we find a different kind of enlightenment from the alienated, the philosophically and even politically radical, version thrown up by the various *ancien régimes* on the Continent. The English "philosophe,"* whether as Fellow of the Royal Society, scientific experimenter, lecturer, or engineer, managed to flourish in the Whig and Erastian† political order that dominated eighteenth-century England. Although it did produce its share of alienated Tory‡ wits, that order fostered a unique intellectual movement centered on the new science, on its cultivation and promotion. In the hands of these scientific philosophes the mechanical philosophy was grafted onto the interests of its audience in a way that helped to lay the foundations of an industrial mentality. This fusion rested on a vision of the profits and improvements made possible by science. Its ultimate success was also contingent on an intelligentsia sufficiently content, at least in the early decades of the century, with the larger political and ecclesiastical order that they were able to concentrate their energies not on attacking the foundations of established authority but rather on rendering scientific learning into mechanisms fit for gentlemen and entrepreneurs.

If the character of eighteenth-century English science is examined closely we find that it was neither moribund nor impractical, although it may not have been as theoretically innovative as some historians would like it to have been. Yet once perceived as dynamic and progressive in relation to the material order, the science of the English philosophes appears not only as a unique version of Enlightenment but also as the historical link between

*See glossary of terms.
†See glossary of terms.
‡See glossary of terms, under Whigs.

the Scientific Revolution in its final, English phase and the cultural origins of the Industrial Revolution.

Unlike their Continental counterparts, the philosophes of the English Enlightenment generally did not have to do battle against powerful and entrenched elites frequently hostile to innovation or to education conducted outside clerical supervision. The English promoters of the new science could concentrate their energies on promoting a version of scientific learning tied to industrial application that their audiences eagerly embraced. Yet this does not mean that these promoters of improvement could not at moments experience an alienation from the existing order comparable to that experienced by Continental philosophes. Late in the century and only within select circles, the English promoters of scientific improvement turned their zeal against the established social and political order. The radicalism of those circles rivals in intensity that found on the Continent during the revolutions of the late 1780s and 1790s. It also harks back to the zeal for reform through the application of science witnessed during the 1640s (see pp. 76–78).

Those circles of radical scientists, generally found in northern England during the later decades of the eighteenth century, participated in a cultural milieu characterized by scientific societies, popular lectures, and books of applied science intended for self-instruction. That culture had first emerged in the early 1700s in London and the southern counties. There we find literate gentlemen, some aristocrats, and Newtonian explicators congregated in coffeehouses or private clubs, where they took up science as part of a larger project of self-improvement and education. Occasionally, some of these gentlemen and their scientific friends joined together in projects to promote the application of mechanical devices to industry, in particular steam engines for mine draining, water pumping for commercial exploitation, or the drying of various substances from hops to gunpowder through the application of steam heat.[4] In effect their understanding of the usefulness of science should be seen as essentially industrial, and it preceded the first Industrial Revolution by well over forty years. Very little came of those projects. Capital was scarce, the engines were faulty, stockholding companies were outlawed after a spectacular collapse in 1720, raw materials were too expensive. The point is that new historical research has shown that the link between Newtonian science and its industrial application was not casual or contingent on purely material conditions for its articulation. The applications had been promoted by Newton's earliest followers long before all the pieces in the material base of industrialization were clearly in place.

SCIENCE AND THE INDUSTRIAL REVOLUTION

What this evidence does is expose as ahistorical the assumption that the "pure science" of Boyle and Newton had nothing to do with the Industrial Revolution.[5] That assumption, of course, anachronistically presumes the existence in the eighteenth century of what can only be described as our own conception of "pure science." As we now know, for many eighteenth-century scientists science was, perhaps above all else, useful science. This was true early in the century, while by the 1790s the linkage between scientific knowledge and industrial application had become commonplace. Indeed by that time the scientific knowledge of applied mechanics may have proved determining when decisions involving the introduction of new machinery, at considerable capital risk, had to be taken promptly and confidently (see Chapter 7).[6] By the 1790s we can find merchants who were able to correct the complex drawing plans of hired engineers. They were able to do so because two or more generations of scientific educators had plied their trade from the London coffeehouses to the valleys of Derbyshire.

The early eighteenth-century Newtonians rendered their science comprehensible to an audience that could be either genteel and educated or commercial and practical. Whatever their social standing or occupation, or lack of the need for one, members of that audience were invariably nonmathematical. In the period from 1691 (the date of the first Boyle lectures) until late in Queen Anne's reign (d. 1714), the effort to reach this audience largely took the form of pulpit lectures, of which the great Boyle lectures given in London churches by liberal Anglican clergy such as Bentley, Clarke, Whiston, and Derham are justly the most famous.[7] By 1710, however, scientific promoters had also found entirely secular milieus for their lectures, which quickly turned into structured courses now offered in coffeehouses, taverns, and publishers' shops. Some of the earliest of those given by William Whiston and Benjamin Worster illustrated the Newtonian universe by recourse to mechanical devices, through demonstrations using weights, pulleys, and levers.[8]

In this genre of early scientific lectures two characteristics are immediately evident and remain present throughout much of the century. First, natural philosophical language, such as we find in the Boyle lectures of Richard Bentley (1692) and Samuel Clarke (1704), continued to be used in the opening lectures of any course, with definitions of matter, motion, space, and time freely tendered, complete with their implications for society and religion. Just as in the Boyle lectures, matter theory—to us that most abstruse of subjects—was routinely explicated. The audience, long steeped in the language of the sermon and prayer book, was made to understand that it is a violation of orthodox Newtonian theory to assert, as would the materialist, that motion is inherent in matter, and thereby to sever the universe from divine control and deny the providential harmony of the existing social order. The consumers of the new science, who might pay anything from one to three pounds for a six-week course that met two or three times a week, were repeatedly told that what they were learning sanctioned the existing social and constitutional order. Second, the earliest lecturers, and all their successors, used mechanical devices of increasing complexity, especially air and water pumps, levers, pulleys, and pendulums, to illustrate the Newtonian laws of motion and hence simultaneously their applicability to business, trade, and industry. It was pointless to give mathematical explications to lay audiences; that must have been obvious from the beginning. The interests of men who wished to weigh and move goods, to improve water transportation, to drain fens or remove the damp from mines dictated the format of the earliest lectures. To that extent the practical interests and mathematical limitations of the audience for science profoundly shaped its articulation; at the same time the experimental rigor of the Newtonian achievement disciplined and excited the minds of its receivers. Through these scientific lectures nature was rendered knowable; its laws could be mastered and, just as important, applied.

Surviving outlines and course descriptions, as well as published textbooks from early to late in the eighteenth century, make this mechanical and applied approach to Newtonian science absolutely clear. Among the most important and early proponents of this method of scientific education was the official experimenter of the Royal Society of London, French Huguenot refugee and Freemason Jean T. Desaguliers. In December 1713 he offered a course of twenty-one lectures on Newtonian science for two guineas wherein he endeavored "to make myself understood to such as are altogether unskilled in Mathematics."[9] In his lectures Desaguliers began quite theoretically—for example, with "an experiment to show what Cartesius meant by his three elements" and with the basic definitions of Newtonian natural philosophical principles: "Matter is what has extension and resistance which are properties of all bodies . . . gravity is an universal principle in Matter . . . the Law of that Gravity, or general attraction . . . decreases as you recede from the center of the attracting body, just as the square of the distances increases."[10] He explicated the Newtonian concept of the vacuum and offered experimental proof for its existence. Yet in the same lecture he moved to the principles behind mechanical engines: "The whole effect of mechanical engines, to sustain great weights with a small power, is produced by diminishing the velocity of the weight to be raised, and increasing that of the power in a reciprocal proportion, of the two weights, and their velocities"[11] In another lecture he proceeded "to show the effect of mechanical engines in general."[12] But the course only began there. General mechanical principles were applied to the operations of levers, weights, pulleys, and the use of wedges. The laws of motion were applied routinely to cannons and bullets—Desaguliers' interest in mechanizing and rationalizing war was lifelong[13]—as well as to the motions of the planets. Specific attention was given to using mechanical principles in order to augment human strength,[14] and to applying mechanical principles to water flow and control. Savery's steam engine figured in these early lectures, and later so did Thomas Newcomen's improvements on it.

Although Desaguliers began his lectures in the coffeehouses of London, he quickly took to the provinces—lecturing, for example, to the gentlemen of the first provincial scientific and literary society, the Spalding Gentlemen's Society, in the small market town of Spalding, Lincolnshire. There he illustrated Newton's three laws of motion, "exploded" Descartes's theory of the vortices, gave a demonstration of "a model of the engine for raising water by fire"—one probably based on Newcomen's steam engine, in which Desaguliers had a great interest—and explained in the section on levers and pulleys how "men or horses of unequal strength may be made to carry, or draw a burden, in proportion to their Strength."[15] Similar lectures were given at the Spalding society by John Booth, who charged twenty persons only half a guinea each to hear about "the universal properties of matter" and "concerning motion in general." The same topics were discussed by William Griffis, another itinerant lecturer who was all over the Midlands in this early period.[16]

Desaguliers also published the extremely influential *A Course of Experimental Philosophy* (London, 1734–1744; translations in Dutch and French), which proclaimed the goal "to make art and nature subservient to the necessities of life" and which used "machines . . . to explain and prove experimentally what . . . Newton has demonstrated mathematically." Once again these printed lectures, complete with illustrative engravings, explicated the law of universal gravitation, the use of scales, levers, pulleys, the bucket engine for raising water, indeed engines of every kind to lessen the need for human labor—in short, every conceivable mechanical device capable of application to industry.

The economic vision of the text is surprisingly consonant with the basic principles around which the Industrial Revolution later occurred. Desaguliers states it very simply in his discussion of the application to coal mining of Savery's steam engine: human labor is expensive, horse power is cheaper but still "a very expensive way"; what is needed is "a philosopher to come, and find a means to bring down the end of the beam [of the water pump] without men or horses" (*A Course of Experimental Philosophy*, vol. 2, p. 468). By the mid-eighteenth century there were many such practical "philosophers" who could understand the principles first explicated by Desaguliers and his associates.

THE SPREAD OF SCIENTIFIC EDUCATION

In the 1720s Benjamin Worster's London lectures were tailored "for qualifying young gentlemen for business," and he attacked those clergy who still opposed the new science as men "whose chief Merit and Trade it is to lye for God."[17] In the 1730s Isaac Thompson, a lecturer of Quaker background, gave a course in Newcastle-upon-Tyne on the mechanical philosophy specifically intended for "those in the coal trade." He repeated it a few years later because of the large number of coal owners of the river Tyne who had subscribed.[18] Both John Horsley (1685–1732) and Benjamin Martin (1704–1782), itinerant lecturers in the early and middle decades of the century, illustrated the Newtonian universe by constant reference to mechanics, going from "the method of computing the force of all sorts of engines" to the application of Newtonian physics to clocks and guns. Martin was so zealous in his desire to convert Newtonian science into universal practice that he lectured publicly to mass audiences. He also came to despise the genteel exclusivity of the Royal Society—at least that was his attitude after he was denied admittance.[19] By the 1760s itinerant scientific lecturing was everywhere in fashion and Martin, although able to make a living at it, had dozens of competitors.

Among the earliest secular occasions where science revealed its mysteries was in the new speculative masonic lodges comprised largely of tradesmen but also frequented by gentlemen and aristocrats. Adulation of Newtonian science was an official part of masonic belief, as revealed in the 1723 *Constitutions* published by the Grand Lodge of London (see pp. 126–127)—a document in which Desaguliers had a considerable hand—and masonic rhetoric reflected quite early in the century a new enlightened definition of the gentleman.[20] In one of the most important changes wrought by the Enlightenment in England, he was now defined as a man of science. Many institutions and trends—not least the provincial scientific societies, to which we shall shortly turn—wrought this transformation, but we can find evidence of it quite easily and early in masonic literature. Lecturing to his brothers in York a masonic orator of 1726 gloried in the world of ordinary mortals: "Human society, Gentlemen, is one of the greatest blessings of life . . . for 'tis to it we owe all Arts and Sciences whatsoever." He condemned the "learned pedant" as an "unsociable animal . . . who has shut himself up all his Life with Plato and Aristotle." The tradesmen present in the lodge were exhorted by the orator to be faithful to their callings; but from gentlemen more was expected: "The education of most of you has been noble, if an academical one may be call'd so; and I doubt not but your improvements in literature are equal to it." Freemasonry asks, however, not only that gentlemen "by signs, words, and tokens . . . [be] put upon a level with the meanest brother," but also that they "exceed them, as far as a superior genius and education will conduct you. I am credibly inform'd [the orator continued] that in most lodges in London, and several other parts of this kingdom, a lecture on some point of geometry or architecture is given at every meeting."[21] There is good reason to believe that lodges in London practised what the York orator preached. At the Old King's Arms Lodge in the 1730s an itinerant lecturer, avid Freemason, and progressive schoolmaster named Martin Clare lectured "on the history of automata . . . on the circulation of the blood . . . [and] on magnetism."[22] Other evidence suggests that Clare also gave his complex lectures on hydrostatics first to his masonic brethren. In those lectures on the motion of fluids he drew attention to the steam engine but issued a caution that would determine its selected use until well into the nineteenth century: "in point of profit" the engine may not answer the expectation of those who use it "where Fewel is not very cheap."[23]

Where we find the association of science with Freemasonry, or with societies whose ambience resembled the relaxed socializing of the lodge, as did the Derby Philosophical Society of the 1780s and 1790s, then we see most clearly the democratizing tendencies within eighteenth-century scientific culture. The Freemasons sought (though only in their private gatherings) "to meet upon the level," and in that spirit Martin Clare attempted to bring science to the lower middle class. His educational exercise book for young apprentices, which was an eighteenth-century bestseller and went through ten editions, made passing reference to the value of experimental and natural philosophy as early as its first edition (1720); by its fifth edition (1740) it actually gave the three Newtonian laws of motion and exercises to illustrate them.[24]

Although we do not as yet associate the masonic lodges of eighteenth-century England with the spread of scientific learning, the historiography of the period takes for granted the important role of the Protestant Dissenters* and their academies in the enterprise of scientific education. The evidence is quite convinc-

*See glossary of terms.

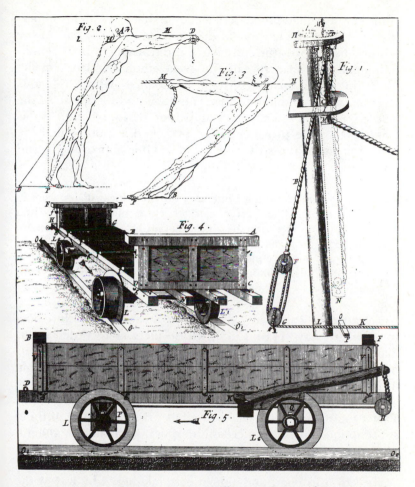

The illustration is from J. T. Desaguliers, *A Course of Experimental Philosophy* (London, 1745), vol. 1, pp. 283–285, showing new carriages used in coal mines near Newcastle and the relative ease with which they can be manipulated by human labor. (*Courtesy of the New York Public Library*)

ing for this truism, and it is a necessary and useful one provided it does not obscure the larger, more widespread dissemination of science in Anglican or purely secular settings. Certainly the inventor of the steam engine, Thomas Newcomen (1663–1729),[25] was a devout Baptist, while Philip Doddridge, the Presbyterian minister of Northampton, was a leader in that city's philosophical society. One of the few extant diaries from the early part of the century that records the conscience and learning of the scientifically minded layman reveals the great emphasis placed in the Dissenting tradition on constant attendance at both religious sermons and scientific lectures. The habit and discipline of the first led, it would seem, to the cultivation of the second. As this diary reveals, the providentialism preached at both suited the temperament of the striver and seeker after improvement and salvation:

> This Day in the Morning I attended on Domestick Affairs, went in the Afternoon to Manchester heard a Lecture concerning Attraction and Repulsion of Matter by Mr Rotheram D.D. [Caleb Rotheram, D.D. (1694–1752), minister at Market Place Chapel, Kendal] we subscrib'd our Guineas a piece each. Lord, May this and all other labor and expense for my improvement and advantage turn to a very good Account.[26]

This listener at Rotheram's lectures was a young Lancashire doctor, Richard Kay (1716–1751), whose piety as recorded in his diary came to embrace scientific instruction as a singular means by which "may I daily grow in Wisdom and Knowledge both in Temporals and Spirituals."[27] His almost daily entries attest to the abundance of lectures already available in the shire by the late 1730s and to the theoretical as well as practical nature of these subscription courses in mechanics, optics, hydrostatics, geography, electricity, and pneumatics.

Fortunately a manuscript copy of one set of lectures transcribed by a listener, and almost certainly heard by Kay at the Angel Inn in Manchester, has been preserved. These "Observations and Memorandums of the Philosophical Experiments in a Course of Lectures; begun August 15, 1743 . . . by Mr John Rotheram, Jr of Kendall" are notes taken from memory by someone present at lectures given by Caleb Rotheram's son, John, and they are extremely useful for illustrating both what was actually said, as opposed to what was printed, and better still, what could be absorbed by the careful listener. Physics, theology, and Newtonian science were effortlessly combined with mechanical devices and experiments to illustrate a wide range of natural phenomena. At every turn it was emphasized "how the all Wise Ruler and Governour of the World has given particular rules and laws to all bodys of every sort and kind whatsoever,"[28] and the lecturer also made frequent reference to Newton's published works, to ideas or experiments given by the Royal Society, and to Newtonian explications by Desaguliers and Whiston. The person recording these lectures is doing so from memory (fol. 45 v), and whenever possible he (or just conceivably she) eschews mathematical illustrations "as mathematical experiments are strange to me."[29] The lecture on motion or gravity quickly moves to a discussion of weight and velocity, explaining that the "very principle and foundation stone on which depends all the Laws of Mechanicks" is the relationship between weight and velocity: "Where a weight of 6 pounds is to be balanced by another of 3 it will require twice the velocity as twice three is six to bring it to an equilibrio." The rest of the lecture concerns beams and the brachia of a pendulum; the audience is even treated to an explanation of how goldsmiths and shopkeepers might fiddle their scales to deceive their customers. Pullies and inclined planes are also examined, with the admonition that "compound machines are the same in effect as the simple machines with this difference only, that simple machines are only smaller in their powers and in their weights than the compound machines and therefore less force is required." The measuring of the force of projectiles for "bombarding and cannonading" is explained, with the method currently in use among military engineers criticized as "not mathematically true." A pyrometer for measuring the heat needed to expand metals is demonstrated, and an entire lecture is devoted to "how fluids gravitate upon one another, and from other experiments to show probable causes for the arising of some phenomena such as the ebbing and flowing wells . . . in Yorkshire and . . . in Derbyshire."[30] A generation later engineers in Derbyshire were to attempt to harness water power to windmills and to install steam engines where cheap coal was available. Where else would they and the entrepreneurs who paid them have learned of the technical risks involved in these mechanical operations except by virtue of the scientific knowledge first transmitted in lectures such as these?

By the second half of the eighteenth century, however, this new scientific culture had permeated so deeply into the lives of the nominally Anglican and genteel elite that it would be quite mistaken to associate it primarily and exclusively with the Dissenters and their academies. In the early part of the century the Dissenters were inordinately active, in relation to their actual numbers, in the enterprise of disseminating science to the commercial and the industrious. Richard Kay's knowledge of the new science was undoubtedly superior to that of a contemporary Cambridge or Oxford graduate, institutions from which Kay was excluded because of his religion. Yet even in the traditional universities some evidence can be found from early in the century of scientific lecturing intended for industrial application. In 1716 Thomas Whiteside, the keeper of the Ashmolean Museum, gave a course of lectures described as excellent by contemporaries in Oxford, and the existence of a copy of these lectures in Cambridge implies that they may have been given there as well. "A Course of Philosophical Lectures" deals almost entirely with mechanics, specifically with the assertion that:

> There is a universal law to which all the forces of mechanical powers may be reduc'd (viz) the power and ye Burden are reciprocally proportional to ye velocities. This is evident in all kinds of levers. Now this Law is applicable to all ye other mechanical powers since they are reducable to the lever. The whole effect then of mechanical engines consists in diminishing the velocity of the weight to be raised, so that its momentum be no more than the momentum of the power that raises it.

Whiteside concentrated on levers, pulleys, inclined planes, pendulums, and screws. But it should be added that his lectures, at least in Oxford, required a subscription of one and a half guineas and may have been intended for a wider audience than simply undergraduates.[31] Other scientific lectures given at Cambridge on Newtonian science appear to have been more overtly theoretical but no less clear or polished.[32] Yet at mid-century Oxford offered little hospitality to Dr. Nathan Alcock (1707–1779) who had gone to Leiden to learn his Newtonian mechanics from s'Gravesande. Returning to Oxford, where in his opinion "little or nothing" was done for the scientific education of the students, he managed to secure a lecturing position "against all opposition."[33] But the deficiencies of the universities could be compensated for by attendance at public lectures, membership in the Royal Society or one of the provincial philosophical societies, or simply by reading one of the many new scientific books then pouring from the presses.[34]

By the last quarter of the century a new ideal of the English gentleman had been proposed and ratified by polite society, and foreign writers grasped it as a standard to which their audience, envious of English prosperity, should now aspire. As the Dutch became increasingly alarmed by the evidence of their own economic decline, they looked obsessively to the English, searching for what it was they were doing "right," in much the same way the English had done a century earlier when they pondered the prosperity of their Dutch rivals. The epistolary novel of the Dutch writer Marie Gertruide de Cambon-van Werken created for her readers both in Dutch and in popular English translations of the 1790s the fictional ideal of the contemporary English gentleman, one Sir Charles Grandison (a name undoubtedly borrowed from Richardson's novel), whose favorite toy as a child was a microscope, with which he studied the insects that crawled about his feet. His tutor, Dr. Bartlett, the veritable English Pangloss, wished to make him wise, "to shew kindness to the insects . . . to let my love mount up from them to the beings, who while they enjoy the blessings of heaven, can recognize the hand which bestows them."[35] Grandison combines science with sentiment and benevolence; he is a true aristocrat of his age, one worthy of emulation for his sense of order, industry, and fair play.

As the popular English translation of this novel proclaimed, Grandison's lessons from Dr. Bartlett rendered him socially tolerant: "I recollected I have often seen labouring men very compassionate. God takes care of the meanest insect."[36] Despite the element here of fictional caricature, not to mention priggishness, it is altogether possible that a real-life Sir Charles would have been eagerly welcomed as a Fellow of the Royal Society or as a member of one of the provincial philosophical societies that appeared as early as 1712 and grew to abundance by the 1750s.

THE SCIENTIFIC SOCIETIES

The itinerant lecturers were the purveyors of scientific culture, but they neither initiated it nor sustained it. For a continuous history of the diffusion of scientific enterprise and learning to the genteel and the mercantile, we must look to the philosophical and literary societies of the period, from the Royal Society of London to the Spalding Gentlemen's Society, the Northampton-shire Philosophical Society, and later in the century, to the Lunar Society and the Derby Philosophical Society, among others.[37] To date we have not adequately emphasized the role of these eighteenth-century private societies in fostering a new enlightened culture. Like its Continental counterpart, the English Enlightenment flourished in the milieu created by secular fraternizing for the purpose of personal improvement and social intercourse. This culture was public and secular, in that it was neither family oriented nor in attendance at the feet of pulpit preachers. Yet it was private in that its elite membership was restricted by income and dues, education and occupation; and the meetings were always held behind closed doors although not necessarily in secret, with the exception of the Freemasons.

As European historians who have attempted to explain the significance of the eighteenth-century philosophical societies for the French Enlightenment have noted, the philosophical society on the Continent prefigured there a new political order, originally English, and one based on parliamentary and constitutional systems that require constant communication within the political nation.[38] Predictably the literary and philosophical societies in England were to some extent an outgrowth of the political and social order produced by the revolutions of the seventeenth century. The first English philosophical society in the provinces began among a group of gentlemen in Spalding who gathered each week to read the Whig journal, the *Tatler,* as it arrived off the London coach. A similar society was started in Edinburgh, also late in Anne's reign, as the outgrowth of a group formed to read a similar journal, the *Spectator.*[39] The free circulation of the printed word was an essential precondition of the new scientific culture. The ferment of party politics within a parliamentary context gave unprecedented circulation and relative freedom to the press. Quite incidentally, that freedom encouraged all sorts of private and public gatherings where ideas and books were discussed. These groups met the social needs of new men who defined themselves as individuals, divorced, at least psychologically, from the old corporate structures of extended kinship, medieval guild, or religious confraternity. For such men, in a minority in any European society of the eighteenth century, the philosophical society composed of individuals dedicated to self-improvement provided identity and conviviality, just as the political club provided an outlet for self-interest and conviction.

We should also bear in mind the importance of surplus wealth, a relatively new phenomenon among the middling classes in western Europe, in permitting these societies to flourish.[40] That element may be determining when we try to understand why certain cities or towns, and not others, might sustain such gatherings. In England among those who possessed excess wealth or, just as important, those who aspired to it, the perception that science would contribute to their aspirations was arrived at quite early in the eighteenth century. As the century wore on, the proportion of time in those societies devoted to science increased, at the expense of history or literature; utility intended for profit became increasingly paramount.

As the most prestigious English philosophical society, the Royal Society of London offered extremely prosperous gentlemen not unlike Sir Charles an intimacy with science unrivaled by any other eighteenth-century forum. All the early provincial philosophical societies recognized that fact and sought copies of its minutes, even before publication, so as to be abreast of the latest scientific advances. But despite the enthusiasm of contemporaries, the Royal Society's dalliance with "amateurs," its insistence on science as useful science, and its gentlemanly membership have cost it dearly with historians of eighteenth-century "pure science." They have simply refused to write its history, lamented its decline, and hinted at its corruption.[41] Although it is not possible here to write that history, some passing reference must be made to the sources out of which it could be constructed.

Throughout most of the century candidates for membership in the Royal Society had to be proposed well in advance by three or more members, who wrote and publicly displayed at meetings "certificates" stating the reasons why a man should be voted in by the Fellows, each of whom possessed one vote. After 1730 the affirmative vote of two-thirds of the Fellows present at a meeting, rather than a vote in the society's Council, was required for admittance. It seems reasonable when reading those extant manuscript certificates from the 1730s to the 1760s,[42] for example, to suppose that they were intended to impress the membership, to reflect the society's best conception of itself and its science in such a way as to court a Fellow's affirmative vote for a candidate. Negative votes were occasionally given, however, and it is also worth speculating why some men and not others were admitted.

The society's rhetorical formulations about itself in the vast majority of cases emphasized the ideal of electing "a gentleman well skill'd in all parts of the mathematicks, natural and experimental philosophy, and most branches of curious and useful learning." But one category was no more important than another. A gentleman from the Navy Office could be skilled in mathematics and have written a treatise on gunnery. Alternatively one could either possess useful learning or simply be "useful"—for example, by defending or explicating one of Newton's works, especially for a foreign audience, or by "being a great promoter of Natural Philosophy," or by supplying the society with foreign contacts as in the case of Mynheer Hop, the Dutch ambassador. It was sufficient for a squire to be "well-versed in natural knowledge" or to be a "great lover of natural knowledge" without possessing any particularly noteworthy scientific achievement.[43]

Philosophical lecturing on the circuit was never held against a man, as Benjamin Martin's rejection might have implied. Martin Clare, who was recommended by his masonic brothers Desaguliers and Ephraim Chambers, was described, obviously with approval, as "a good mathematician well skilled both in natural and experimental philosophy, and a great promoter of the same."[44] Being a mechanic as well as a merchant was also no shame. Jonathan Fawconer, a lapidary in London, was described as having "a knowledge of precious stones [while being] well-versed in most branches of mathematicks, and more particularly in mechanicks, having invented an engine of great service to him in his profession." In 1735 the Society honored a man with election who presented before it "curious experiments relating to the damp of coal mines"; another was admitted because as a sea captain he had made many geographical and navigational inquiries on the Society's behalf. His admittance stands in contrast to the rejection of a doctor of physics and a surgeon, the latter being recommended by Martin Folkes, Hans Sloane, and Jean Desaguliers.[45]

Generally speaking men with such prominent fellows as their sponsors did not suffer rejection, and it is as yet difficult to determine why certain men were turned away when their reputation, occupation, and sponsorship were apparently acceptable. The rejection of the French philosophe Diderot in 1752 should not, however, be surprising. His reputation as a philosophical radical and a materialist would have doomed him. What is fascinating, however, is to see one of the highest turnouts of Fellows for any meeting in the century where a membership vote was to be taken; Diderot was rejected by 50 votes to 18.[46] Obviously there were other criteria for membership beyond those stated in the certificates, and these would eventually lead to dissension within the society.

In the 1780s a rift opened publicly between the president of the Royal Society, Sir Joseph Banks, and Fellows who felt that Banks was using his personal influence excessively, "making himself the sole master of the admissions, in other words, the *Monarch* of the Society." Banks was not trying to restrict membership to the "pure scientists"; he would not have understood the nature of the Society or science in those terms. Rather he was trying to keep out the less socially prominent and those of republican leanings, some of whom were friends of the radical reformer Dr. Priestley, himself a Fellow. Neither were his opponents trying to admit the "pure scientists"; they too would not have accepted the distinction. Instead they argued that "for whatever we ought to be (which is another question), we are not an Academy of Sciences, i.e., a receptacle for the Great in Science, but a society of Gentlemen, of all ranks and professions, all opinions, and, we must add, all kinds of learning (or no-learning) paying 52s. a year for the encouragement of literature."[47] The reformers present what the certificates for admission confirm to have been the Royal Society's conception of itself and the variety of its membership.

The Royal Society's conception of good science tells us much about that British cultural ideal on the eve of industrialization. The propensity to link science with application, with trade and industry, was part of the ideology that created the Royal Society in 1662. It harks back directly to Samuel Hartlib (see pp. 74–76) and the Puritan vision, and hence to Francis Bacon. In the late seventeenth century that industrial impulse led to the attempt by Fellows to compile massive histories of all the known trades, from artillery to masonry, coal mining, and iron foundry. In the early eighteenth century the Society gave its annual Copley awards for successful industrial application: in 1738 to a watchmaker "for his useful engine contrived for the driving of piles of a new bridge"; in 1739 to Stephen Hales "for discovery of dissolvents of the stone, and preservation of flesh in long sea voyages."[48] The pattern of those awards for the rest of the century does not deviate significantly from that criterion of usefulness broadly conceived. In 1759 when the president of the Royal Society, Lord Macclesfield (1697–1764), awarded the Copley Medal to the famous civil engineer John Smeaton, to whom we shall return, he noted that while the medal should be for papers presented to the society "yet much honor" belonged to Smeaton for his extraordinary achievement in constructing the Eddystone lighthouse. Macclesfield in turn praised Smeaton for his papers submitted to the *Philosophical Transactions* on water wheels and windmill sails, for displaying in them "so clear and scientific method" to produce "a chain of close reasoning, that depends upon many computations and numeral proportions arising from the results of a considerable number of very accurate experiments."[49] In short, Macclesfield and the committee that awarded the Copley Medal thought that Smeaton was a very fine scientist. As we shall see from his private letters, Smeaton also thought this of himself, and so did his friends. The definition of science employed by the Royal Society, although laying greater emphasis on the mathematical and experimental than was common in the other philosophical societies of the period, was held in admiration by the gentlemanly members of those societies and never seen as demanding an enterprise that was either intimidating or foreign to their aspirations.

But organizing natural philosophical learning into provincial societies, given London's preponderance and the difficulties in transportation and communication, was seen as a delicate business. The success rate for the early part of the century was not high. Yet given the notable preponderance of those societies in

British, as opposed to Dutch or French, cultural life, we should examine that unique phenomenon in some detail. Indeed one of the most extraordinary examples can be found in the Spalding Gentlemen's Society.[50] From 1712 to 1755, 374 men joined this society located in a town of some 500 families, although only about 20 or so members in every decade formed its core. Its organizer and guiding light, Maurice Johnson, belonged to the top stratum of genteel society, was a Justice of the Peace in the shire, an active Fellow in the Royal Society, and a devotee of London club and pub life. He sought to export that easy socializing to the county, to the gentlemen of quality but also to clergymen, who were to form nearly a fourth of the society's membership, and to doctors, lawyers, surveyors, and even tradesmen, the last of whom made up 3½ per cent of the society's total.[51] The Spalding Gentlemen's Society's manuscript minutes show a remarkable interest in Newtonian science as explicated by s'Gravesande and Desaguliers—provided the explanations were not too mathematical—and in natural artifacts and natural history in general as well as in antiquities.[52] At the inner circle of the society's leadership was a ubiquitous medical man, later turned clergyman and antiquarian with an interest in Freemasonry and the mystical, one Dr. William Stukeley (1687–1765). By his side was John Grundy (1696–1748), one of the most important engineers in the period between Newcomen and Smeaton.[53] Both incidentally were Freemasons; indeed Grundy was master of the lodge at Spalding, one dedicated to the improvement of each brother "through natural philosophy" among other disciplines, and Johnson may also have been a lodge brother.[54]

The Spalding society is important not simply because its very existence demonstrates the early and widespread interest in the new science among the genteel, the educated, and the mercantile of the provinces, but also because its understanding of science, as documented in its archives and extant library catalogues, was mechanical and experimental as well as "useful." In this society the itinerant lecturers made their mark on men interested in the draining of the local fens; in the improvement of agricultural techniques; in surveying for enclosure and canal building; in using pumps, pulleys, levers, and the new steam engines, wherever practical; and in finding labor-saving devices of a mechanical sort. At least one scientific educator of the period, John Rowning, author of *A Compendious System of Natural Philosophy* (1738), regularly attended the society's meetings.

But did this science of the early lecturers, to be found in Spalding as well as in London, ever actually inspire technological innovation of the kind essential to the early stages of industrialization? The records of the Spalding society prove that it did. The water engineer John Grundy, for instance, has always been described as self-taught, and his important innovations as a surveyor of the fens and instigator of various successful canal routes through Lincoln, Chester, and Lancaster have never been connected in the historical literature with his interest in natural philosophical learning. He is known to have taught mathematics in the 1730s and may even have been an itinerant lecturer early in his career; and perhaps most important, he educated his son of the same name (b. 1719) to such a degree of excellence that he became an important engineer of the next generation. But the manuscript minutes of the Spalding society reveal Grundy's interest in the industrial and agricultural application of the new science:

> Mr. Grundy communicated his proposals for draining lands that lie near the sea, showing the necessity of mathematical and philosophical knowledge thereto required approved by J. T. Desaguliers, D.D., F.R.S.[55]

Desaguliers we know had actively campaigned for the industrial application of science in England and the Low Countries.[56] But Grundy had gained more from him than simply an interest in machines; he had learned the principles of Newtonian science as an experimental and mechanical system. When Grundy's machines pumped efficiently, they did so in part because their inventor obtained through the Spalding society a theoretical and correct understanding of the natural order that enabled him to channel nature into the service of human productivity.

Other philosophical societies from this early period also sprang up in Stamford, Northampton, Peterborough (1730), and Edinburgh. But the Peterborough society, started by a clergyman from the Spalding society, Timothy Neve, refused to admit tradesmen, "which makes some who set themselves up for gentlemen the more desirous of becoming members, as piqueing themselves upon their quality or profession.''[57] Although modeled on Spalding, the Peterborough society barely survived ten years. The one at Stamford (not far from either Spalding or Peterborough) did not do much better despite the occasional appearance of an original Newtonian, William Whiston, at its meetings.[58] It would seem that these other societies lacked dynamic leadership. This was the critical factor, since both Northampton and Spalding succeeded for a time and both possessed nationally connected organizers of stature: Johnson in the case of Spalding; and Thomas Yeoman, the engineer, and Philip Doddridge, the Dissenting minister, in the case of Northampton. Yeoman was an itinerant lecturer and manager of a cotton mill in Northampton who turned his skill to surveying for turnpikes and enclosures as well as to water engineering. Indeed in 1743 he erected the machinery in the world's first water-powered cotton mill.[59] In these and other projects he had the encouragement of Doddridge, who extracted from the Newtonian legacy both its mechanical and ethical applications.[60]

More than enthusiasm for Newtonian science was required to create and sustain these societies. All evidence suggests that their membership was drawn entirely from the ranks of voting freeholders—those newly empowered political individuals. Where we can find ideological affiliations for the leadership of these societies, whether in London or Northampton, it was court and Whig in its identification with the ruling oligarchy. The Northampton society, for example, was filled with supporters of the Whig parliamentary candidate whose 1748 election was one of the most hotly contested of the period. The Whig leadership of the Royal Society in this period is, of course, well known. The societies eschewed politics, yet they were dependent on political life for the social network that sustained them. Other social factors, however, were equally important in giving dynamism to these gatherings. As the quick demise of the Peterborough society would suggest, exclusivity if carried too far brought with it boredom. Those societies that permitted the mixing of the genteel and educated with the mercantile and prosperous seem to have kept both their attendance and their intellectual content reasonably high.

What, however, was a man, and certainly a woman, to do if he or she did not have access to, or could not afford to be admitted to, one of these societies? For both there were, of course, the traveling lecturers, who specifically appealed to ladies to attend their courses, occasionally even offering a special course solely for them. Some lecturers preferred daylight for their demonstrations and therefore needed leisured customers to pay the fees as well as to attend these afternoon sessions. After all, a course in science at two guineas would be no more than a gentleman would pay for his children to attend lessons with a dancing master. Aside from these public lectures a woman could also read about science, and by mid-century the finest scientific journal for a lay audience was the *Ladies' Diary.* It set complex mathematical puzzles for its female readership, but copies of solutions offered by members of the Spalding society turn up in its archives.[61]

JOHN SMEATON

One such lone reader and student of the new science, who became a close friend and admirer of the younger John Grundy, was the Yorkshire engineer John Smeaton, F.R.S. (1724–1792). With him we move from the general cultural background, the intellectual origins of the Industrial Revolution, into the very center of that economic and social upheaval. Recognizing as we do now that water power was more important in the first decades of industrialization than steam, Smeaton must be seen as singularly important for his perfecting of the atmospheric engine—indeed, as the most important industrial engineer in the years between Newcomen and Watt.[62] Like the older Grundy, he has been described in the literature about early industrial technology not only as "self-taught" but also as being without "any system of conceptual or theoretical knowledge."[63] The evidence proves otherwise.

Like so many other minor, or major, philosophes, Smeaton's earliest intellectual interests began with religion, as his unpublished letters to Benjamin Wilson testify. Wilson, who was an early electrical experimenter and theorist, acted for a time as Smeaton's mentor in matters religious and scientific. Both, it would seem, sought a rational Christianity; and in the context of either eighteenth-century Anglicanism or Dissent that meant discarding the doctrine of the Trinity. Smeaton believed that "this is doing service to religion . . . Jesus Christ [is] the greatest of all created beings and mediator between God and man . . . I think it is observable there is no direct mention of a Trinity . . . in the Scripture."[64] Throughout his life Smeaton paid little attention to formal religiosity, maintained his belief in divine providence, and regarded the Catholicism he witnessed on the Continent as "calculated for striking the minds of the vulgar."[65] Indeed other people's prejudices and superstitions fascinated him, while he and his engineer friends prided themselves on their willingness to follow wherever the new science led, even in the dangerous business of allowing their children to be inoculated against smallpox.[66] As a young man Smeaton sought to become a "philosopher," to use his word, and he cultivated a sophisticated understanding of Newtonian science. In his letters he offered Wilson both theoretical and experimental criticism for his ideas on electricity, but he also believed that learning must serve society:

> If I were so far to abstract myself from the world as to consider myself without either friends or enemies or any person to take any measure of notice of me . . . I should quickly lay aside toyling and noyling in ye sciences, and industry of all kinds any farther than so much food as to provide me from starving.[67]

Smeaton, like so many enlightened men, believed that the love of society was implanted in every man and that knowledge must serve human needs and bear reference "to the opinion of [our] fellow creatures." Yet amid this idealism reality dawned harshly on this would-be philosopher. His parents objected successfully to his betrothal to an impecunious lady, and around the same time Smeaton realized that even a coveted membership in the Royal Society cost a great deal of money. Its admittance fee of 23 pounds led Smeaton to complain bitterly, "I think only rich philosophers can afford to pay: so that I suppose ye popish proverb will fit them 'no penny no pater noster.' " Although "we ought all of us to be philosophers," this young scientist who wanted to become a professor of physics became instead an engineer. His earliest efforts at building mechanical devices—for example, an air pump with which to create a better vacuum—had been prompted by their usefulness in electrical experimentation.

Smeaton was, however, "as poor as a churchmouse,"[67] and he was forced to abandon the purely theoretical and experimental. He was to struggle for many years to build canals, bridges, and machines for industrial application. Throughout his mature years he sought to attain solvency and respectability, as would any rising professional in a world obviously dominated by the well-born and the genteel. "My profession is as perfectly personal as that of a Physician or counciller at Law," he declared to a local schoolmaster seeking to place a pupil with him, while to a customer Smeaton tersely explained, "The construction of Mills, as to their Power, is not with me a Matter of Opinion it is a Matter of Calculation."[68]

Throughout his life Smeaton's friends frequently addressed him as "my great philosopher"; but how did he perceive himself? Not only did Smeaton invent the profession of the civil engineer —indeed he and the younger Grundy corresponded with one another as early as 1764 as members "of a profession"[69]—he also possessed a finely honed definition of his place in the scientific world, one with overtones of the supposed split between technicians and "pure scientists." But in his own mind, Smeaton is the theorist, the "pure" scientist. In a long letter to a friend prompted by the failure of the London projectors to compensate promptly and adequately the architects and surveyors who designed Blackfriar's Bridge, Smeaton wrote of such designers as "artists" who must operate in an economic climate fraught with dangers. Smeaton accepted the laws of the marketplace—"profit . . . is the basis of trade and commerce"—but he objected to the unfair practices of the profiteers who failed to compensate engineers for their time and labor.

The point here is that Smeaton, like Wedgwood, the Strutts, and countless other entrepreneurs of the early Industrial Revolution, believed that the application of scientific knowledge within the free market should be unencumbered and justly rewarded. Indeed in his private letters Smeaton could take up the language used by contemporary political radicals who sought to rally the middling classes against the oligarchy. He speaks of "liberties and properties" as being the "time and skill of any artist employed in design." Such artists by their "superior ingenuity" furnish "employment for thousands" and as such deserve the protection of the government. British entrepreneurs were prepared to turn against that government only when they saw their dreams of social and economic improvement based on science and industry actually endangered.[70]

Smeaton, the scientist, who had also to be an entrepreneur if he was to survive, could identify—if only momentarily—with radical reformers when the dignity of his profession was maligned and his right to profit thwarted by his far richer employers. Generally, however, Smeaton directed his entrepreneurial energies into the discovery of labor-saving mechanical devices that could replace human labor. He, even more than his scientific predecessors such as Desaguliers, perceived that replacing human power with mechanical power increased profits. As long as he could keep his dignity, exercise his talents, and profit, Smeaton never railed publicly or even privately against the existing order. He was too busy being one of the most successful entrepreneurs and scientists of his generation.

RADICAL TENDENCIES WITHIN INDUSTRIAL CULTURE

The Whig and Erastian* order that enabled such men as Smeaton, Grundy, Desaguliers, and their associates to pursue science for its profitable application to industry rendered them at most times politically complacent. In general the historian searches their meeting books or private letters in vain for a hint of political disaffection from the ruling oligarchy, for an attack on "corruption" or the evils of placemen or paper credit.[71] These are supporters of the oligarcy by default. When alienation does appear within the mainstream of the organized scientific community, it comes from men of merit rather than birth, who, like Smeaton, think that their skill and talent have not been sufficiently appreciated and rewarded. But what if this scientific culture of the philosophical societies and engineers, with its extraordinary sense of the profit and improvement to be achieved through study and experimentation, were to turn its energies in a socially reforming direction? What if the pursuit of profit through mechanical ingenuity were put in the service of opposition to exist-

*See glossary of terms.

ing authority for the purpose of republican, even democratic, reform? Such opposition among scientific reformers can be found on the Continent in the 1780s (see pp. 195–198); it was rare in eighteenth-century Britain, but it did occur. The dynamism produced by the vision of improvement that the new science inspired could threaten those elements within elite culture who failed to embrace it. Yet the alternative vision of the reformers, so wedded as it was to capital and industry, also possessed distinct limitations.

No general portrait of English scientific culture in the eighteenth century would be complete without discussing the vision and limitations of its radical underside. Public education through lectures, books, and philosophical societies intended to make the genteel practical and the meritorious genteel—with all the optimism that those activities imply—could and did have leveling tendencies, even if most of their earliest propagators self-consciously sought to reinforce the existing social hierarchy. The English republic of letters could also be republican even, or perhaps more especially, when it was busy being entrepreneurial and industrial.

Of the many philosophical societies that sprang up in the later decades of the century, the one at Derby in the northern county of Derbyshire deserves our special attention for what it can tell us about the potentiality and perimeters of enlightened scientific culture. In that society, established in 1784 by Erasmus Darwin (the grandfather of Charles) and intended for the pursuit of "gentlemanlike facts," we find the same useful and mechanical science practised in Spalding, London, or Birmingham, as well as the same desire for industrial application. Now, however, there are industrialists present in goodly numbers. Suddenly the whole burden of scientific culture has also been shifted in a politically and socially radical direction by the addition of two ideological ingredients: philosophical materialism and republicanism. Since the late seventeenth century both could be grafted quite easily onto the new science (see pp. 117–120)—however hard the Newtonians had worked to prevent the hybrid—and now in the Derwent Valley, at the very center of the Industrial Revolution, we find pioneering industrialists like William Strutt (d. 1830), Erasmus Darwin's closest friend and founder member of their philosophical society, distributing copies of the American revolutionary Thomas Paine's *Rights of Man* to his factory workers. Indeed Derby's radical corresponding society of 1791[72] was a direct outgrowth of its philosophical society, and revolutionary ardor, first of the American and then of the French variety, ran through Darwin's society; among some of its members it even survived the disillusionment brought on by the Reign of Terror.[73]

In the Derby Philosophical Society many strands found in the older scientific culture of the mid-seventeenth century are once again present, only now rewoven into an entirely different tapestry. As might be predicted, the philosophical society has become a testing ground for democratic ideas, and even a center for the expression of revolutionary ideals. There still remains the omnipresent emphasis on mechanisms and utility. It is revealed in the extant catalogue of the society's carefully chosen scientific library. This improving science has also been channeled into creating mechanisms for domestic comfort and home improvement, as well as into town planning and the construction of more efficient factories. Dissenters and Unitarians* are also present in exceptionally large numbers, although the irreligious, such as Darwin, seem particularly comfortable among the inner coterie of the society's members. They are devoted to books and print culture in general, collecting as many of the proceedings of scientific societies from anywhere in the world that they can lay their hands on.[74] And finally there are the medical men, as well as the engineers and industrialists, some of whom, like Strutt, are so idealistic and confident about the future being prepared by science and industry that they are ready to promote republican ideas among their workers.

Erasmus Darwin, more a philosopher than an organizer, nevertheless deserves to be seen as the guiding force in Derby's philosophical society. He was a full-blown pantheist and materialist devoted to drawing from the new scientific culture of his age all of its infinite possibilities. And therein lies the root of his political radicalism. His passion for scientific learning, born out of his conviction that "Nature is All," gave him an optimism about human nature and the future that demanded the transformation of existing social and political institutions. His faith in material progress wedded to an egalitarian ideal, although un-

*See glossary of terms.

doubtedly born of a more radically materialist metaphysics, is remarkably similar to that of his near contemporary Thomas Jefferson. Jefferson, however, would act out his ideals in the more congenial political context found in the revolutionary American colonies.[75] Indeed Darwin was on close personal terms with Dr. William Small (b. 1734), who had taught natural philosophy and mathematics to Jefferson at the College of William and Mary in colonial Virginia before returning to Birmingham in 1765, where he, Matthew Boulton the industrialist, James Watt of steam engine fame, and Darwin formed their own *collegium.*[76]

In his opening address to the Derby Philosophical Society, Darwin waxed eloquent about the material and intellectual benefits promised and already delivered by science, which the printed word has "scattered among the great mass of mankind the happy contagion of science and of truth."[77] In a letter to a friend he compared the new society to Freemasonry, as simply another form of congenial fraternizing; but Darwin does not appear to have been drawn to that particular form of secretive philosophical socializing. Nor were other members of the Derby society—only one, a Dr. Pigott, was active in the local lodge.[78] But Darwin and his society were well acquainted with other aspects of the radical tradition; they were republicans of a kind. Among them we find no ambivalence, however, toward capitalism. While that ambivalence toward commerce was undoubtedly present in the writings of many early eighteenth-century English republicans, it is absent from the thought, and most especially from the actions, of Darwin and his friends. Science could work profound transformations within the radical tradition. Inspired by a powerful faith born of science, and reinforced by the wonders of industrial productivity, these Derbyshire republicans put their trust in capital harnessed to machinery, indeed in mechanisms of every sort.

But faith was not enough to save institutions, whether traditional or radical, from experiencing a difficult time in the period of industrialization after 1790. Not only did the masonic lodges in Derby decline, so too did the Anglican churches in the shire.[79] The new population drawn to the district in search of factory work had little use for either institution. They also seem to have had little faith in Strutt—his books or his projects; and his schemes for enclosing the town green, enacted through a parliamentary bill of 1792, were violently attacked by ordinary folk as another form of enclosure.[80] In that decade the philosophical society also came in for severe criticism; the Anglican clergy attacked it and the Dissenters for their support of the French Revolution, for their supposed Jacobinism.* All the reforming and philosophical societies, in particular the corresponding societies, were attacked as followers of the republicans of the 1650s, James Harrington and Milton, as well as of Algernon Sydney and Locke.[81]

In the face of severe local opposition to their politics, the members of the Derby society simply turned their attention away from political reform and busied themselves in improving their town, its factories, its canals, even its homes. In the 1790s these Derby republicans concentrated their reforming instincts on canal building, on the protection of their industrial secrets from foreign spying and economic competition from Irish manufacturers, on public works such as town lighting and domestic improvements such as central heating and the indoor toilet.[82] They talked of projects to build the ideal factory, with a central observation point from which all workshops and workers could be overseen. No amount of popular or clerical opposition could deter Strutt and his philosophical allies from the pursuit of material progress and profit based on science. In the end that revolution took precedence over all others. Although personal letters, particularly from the female members of the Strutt and Darwin families, bear witness to the survival of republicanism, utopianism, and irreligion within the family,[83] these receded increasingly into the realm of the private and the domestic.

The pursuit of scientific learning and its industrial application also continued uninterrupted at the meetings of the Derby Philosophical Society. Nothing could shake its faith in the improvements made possible by science, even if these now had to be confined entirely to the material order. By 1820 the mechanisms championed by the society and its members had indeed conspired with capital and cheap surplus labor to transform Derbyshire in ways that the mechanists of earlier generations could never have imagined. What had begun in the London coffeehouses and taverns during the early 1700s, and then been spread

*See glossary of terms.

by itinerant lecturers and philosophical societies, had finally produced a new kind of entrepreneurial and philosophical gentleman. This industrialist championed a particular type of science, which had to be applied mechanically in order to be understood and which as a result had within its power the capacity to transform both nature and society.

The radicalism of this new industrial culture, where it occurred, centered on the destruction of those landed interests that inhibited its progress. These early industrialists, among them Josiah Wedgwood, came to see the old landed elite as largely hostile to their industrial projects.[84] They extolled the power of the new science, using a metaphor drawn from electrical experiments, as capable of blasting "the oppressors of the poor and needy," of executing "some public piece of justice in the most tremendous and conspicuous manner, that shall make the great ones of the earth tremble."[85] When they were not busy building their industries, they worked for the election of radical Whigs to Parliament and read the work of the French philosophes, not least the democratic writings of Rousseau.[86] The reforming vision of such industrialists presumed constant material improvement intended for the general good. It also presumed that industry and capital would remain in the hands of their rightful owners. As Wedgwood put it, the workmen are "our inferiors"; yet, he conceded, they are made of the same stuff as ourselves and "are capable of feeling pain, or pleasure, nearly in the same manner as their Masters."[87] Part of that presumed superiority derived from the power conferred on the literate elite by scientific education. By the end of the century it was simply assumed that the mechanization of manufacturing, and hence of labor, required a working knowledge of applied Newtonian science— that is, "the laws of mechanics as a science," "the laws of hydraulics and hydrostatics," "the doctrine of heat and cold." With those, manufacturers could mechanize their factories through the application of steam power.[88]

The scientific education of the workers, or "mechanics" as they were called, could be countenanced and encouraged by these masters as a way of inculcating wonder at the rewards promised by orderly diligence and the rational application of mechanical principles. As a result, in the early nineteenth century scientific knowledge finally escaped the small elite within which it had matured. Scientific education gradually became a part of general schooling for boys and girls as well as for adult workers who could attend the "mechanics institutes" that sprang up all over Britain.

Among the first working-class radical intellectuals of the early nineteenth century, useful science, not surprisingly, was to be valued. Once wrested from the pious injunction to obey the master as nature obeys the creator, that science offered a weapon by which the secrets of the machine might be mastered and its benefits channeled to the profit of all mankind, not simply to that of "the Masters."[89] These early worker-intellectuals rightly discerned that the Baconian vision had indeed brought unimagined benefits to those who mastered the methods and application of science. They understood that those who control science and scientific education shall always be its beneficiaries.

NOTES

1. The most concise statement of these mechanistic concepts can be found in Carlo Cipolla, ed., *The Emergence of Industrial Societies,* The Fontana Economic History of Europe (Hassocks, Sussex: Harvester Press, 1976), in particular Phyllis Deane, "The Industrial Revolution in Great Britain." But for evidence of new doubts, see D. C. Coleman, "Proto-Industrialization: A Concept Too Many," *Economic History Review,* vol. 36, no. 3 (1983), pp. 444–445: "And the mere existence of so-called proto-industrialization in such regions was no guarantee whatever of the appearance of entrepreneurial skills or of the capital necessary to induce changes in production techniques." For a particularly mechanistic version of development theory that in order to sustain its case must dismiss the scientific culture described in this chapter, see Ester Boserup, *Population and Technology* (Oxford: Blackwell's, 1981), p. 4. The opening quotations are from *Letters of Josiah Wedgwood, 1762–1772* (London, 1903), pp. 24, 165.
2. For a succinct statement of the thesis, see A. Rupert Hall, "Engineering and the Scientific Revolution," *Technology and Culture,* vol. 2, no. 4 (1961), p. 334: "The great discoveries of mathematical physicists were not merely over the heads of practical engineers and craftsmen; they were useless to them."
3. Both questions are posed most succinctly and provocatively in J. G. A. Pocock, "Post-Puritan England and the Problem of the Enlightenment," in Perez Zagorin, ed., *Culture and Politics: From Puritanism to the Enlightenment* (Los Angeles: University of California Press,

1980), especially pp. 106–108; and Margaret 'Espinasse, "The Decline and Fall of Restoration Science," in Charles Webster, ed., *The Intellectual Revolution of the Seventeenth Century* (London and Boston: Routledge and Kegan Paul, 1974), pp. 347–368 (reprinted from *Past and Present*, no. 14 [1958]). For an interesting perspective on decline, citing the absence of experimental physics at the Royal Society, see J. L. Heilbron, *Physics at the Royal Society During Newton's Presidency* (Los Angeles: Clark Library, 1983).

4. I have benefited from the groundbreaking research of Larry Stewart; see his "The Selling of Newton: Science and Technology in Early Eighteenth Century England," *Journal of British Studies*, vol. 25 (1986), pp. 178–192.

5. Others have exposed the weaknesses in the "pure science" argument; A. E. Musson and Eric Robinson, *Science and Technology in the Industrial Revolution* (Manchester: Manchester University Press, 1969); A. E. Musson, ed., *Science, Technology and Economic Growth in the Eighteenth Century* (London: Methuen, 1972), in particular p. 14, citing Simon Kuznets, *Secular Movements on Production and Prices* (Boston, 1930), for a theoretical background to this approach; Neil MacKendrick, "The Role of Science in the Industrial Revolution: A Study of Josiah Wedgwood as a Scientist and Industrial Chemist," in M. Teich and R. Young, eds., *Changing Perspectives in the History of Science* (London: Heinemann, 1973), pp. 274–319, in particular pp. 274–279, for an excellent introduction to the historiographical problems; and in the same vein, D. S. L. Cardwell, "Science, Technology and Industry," in G. S. Rousseau and Roy Porter, eds., *The Ferment of Knowledge* (Cambridge: Cambridge University Press, 1980), pp. 449–483, with good intuition on Smeaton, p. 470; and finally Peter Mathias, "Who Unbound Prometheus? Science and Technical Change, 1600–1800," in Peter Mathias, ed., *Science and Society 1600–1800* (Cambridge: Cambridge University Press, 1972), pp. 54–80, where I think the problems posed do not bear up against the archival evidence presented by Musson, Robinson, and others. For a recent example of rearguard action from a new version of the "pure science" perspective, see Michael Fores, "Francis Bacon and the Myth of Industrial Science," *History of Technology*, vol. 7, pp. 57–75.

6. Comparative studies of differing rates of industrialization fail almost entirely to speak about the mentality of the elites being compared. Yet there is growing discomfort with this failure and the way in which it has impoverished the entire question of why industrialization might occur in one area of Europe and not another; see Joel Mokyr, "Industrialization in Two Languages," *Economic History Review*, 2nd ser., vol. 34 (1981), pp. 143–149.

7. See Margaret C. Jacob, *The Newtonians and the English Revolution, 1689–1770* (Ithaca, N.Y.: Cornell University Press, 1976); Larry Stewart, "The Structure of Scientific Orthodoxy: Newtonianism and

the Social Support for Science, 1704–1728" (Ph.D. diss., University of Toronto, 1978).

8. Benjamin Worster, *A Compendious and Methodical Account of the Principles of Natural Philosophy: as they are explain'd and illustrated in the Course of Experiments: Perform'd at the Academy in Little Tower Street* (London, 1722); and Francis Hauksbee, *A Course of Mechanical, Optical, Hydrostatical, and Pneumatical Experiments: To be Perform'd by Francis Hauksbee; and the Explanatory Lectures Read by William Whiston* (London, n.d. but probably 1714). The first lecture is on Newton's laws of motion; the second on "the ballance and stillyard . . . All the various Kinds of Levers . . . Pulleys"; the third on the wheel, the wedge, the screw, and "a compound engine." On Whiston, see James E. Force, *William Whiston* (Cambridge: Cambridge University Press, 1985).

9. J. T. Desaguliers, *Physico-Mechanical Lectures. Or, an Account of what is Explained and Demonstrated in the Course of Mechanical and Experimental Philosophy* (London, 1717), preface. See also Larry Stewart, "Public Lectures and Private Patronage in Newtonian England," *Isis*, vol. 77 (1986), pp. 47–58.

10. See British Library, C. 112, fol. 9, *A Collection of Medical Advertisements*, no. 181 (a single-sheet printed course outline). I owe this reference to Peter Wallis; see his paper "Ephemera Issued by the Early Lecturers in Experimental Science," available from the author, University of Newcastle-upon-Tyne.

11. Desaguliers, *Physico-Mechanical Lectures*, pp. 1–5.

12. British Library, C. 112, fol. 9, *A Collection of Medical Advertisements*, no. 181.

13. J. Ozanasm, *A Treatise of Fortification*, trans. and amended by J. T. Desaguliers (Oxford, 1711).

14. Desaguliers, *Physico-Mechanical Lectures*, p. 22.

15. John Theophilus Desaguliers, *A Course of Mechanical and Experimental Philosophy: Whereby anyone, although unskill'd in Mathematical Sciences, May be able to understand all those Phaenomena of Nature . . .* (London, 1725). Preserved in the archives of the Spalding Gentlemen's Society; my thanks to its curator, Mr. Norman Leveritt.

16. John Booth, *Course of Experimental-Philosophy* (n.d.), a flier from the archives of the Spalding society on which are listed the names of members who subscribed; and Will. Griffis, *A Short Account of a Course of Mechanical and Experimental Philosophy and Astronomy* (n.d.; but August 15, 1748, written in as the date the lecturers were first announced to this society) and to be found in its archives.

17. Worster, *The Principles of Natural Philosophy*, preface and p. 230.

18. F. J. G. Robinson, "A Philosophic War: An Episode in Eighteenth-Century Scientific Lecturing in North East England," *Transactions of the Architectural and Archaeological Society of Durham and Northumberland*, vol. 2 (1970), p. 101.

19. See John R. Millburn, *Benjamin Martin, Author, Instrument-Maker, and Country Showman* (Leiden: Noodhoff, 1976), pp. 40–41, 64. In gen-

eral, see F. W. Gibbs, "Itinerant Lecturers in Natural Philosophy," *Ambix*, vol. 6 (1960), pp. 111–117. See also John Horsley, *A Short and General Account of the Most Necessary and Fundamental Principles of Natural Philosophy* (Glasgow, 1743); and Benjamin Martin, *A New and Comprehensive System of Mathematical Institutions Agreeable to the Present State of the Newtonian Mathesis* (London, 1764), vol. 2. Dutch translations of Martin's lectures were bound with those of Desaguliers. See B. Martin, *Philozofische Onderwijzen; of Algemeene Schets der Hedendaagsche Ondervindelyke Natuurkunde* (Amsterdam, 1737). Cf. John Millburn, "The London Evening Courses of Benjamin Martin and James Ferguson, Eighteenth-Century Lecturers on Experimental Philosophy," *Annals of Science*, vol. 40 (1983), pp. 437–455.

20. For a copy of the *Constitutions* and a discussion of Desaguliers's role, see Margaret C. Jacob, *The Radical Enlightenment: Pantheists, Freemasons and Republicans* (London: Allen and Unwin, 1981), appendix and pp. 109–113, 122–127.

21. By a junior grand warden. *A Speech deliver'd to the Worshipful and Ancient Society of Free and Accepted Masons. At a Grand Lodge, Held at Merchant's Hall, in the City of York, on St. John's Day, December 27, 1726* (York, 1726). See in particular pp. 2, 14–15.

22. W. K. Firminger, "The Lectures at the Old King's Arms Lodge," *Ars Quatuor Coronatorum*, vol. 45 (1935), pp. 255–257.

23. Martin Clare, *The Motion of Fluids . . .* (London, 1735); dedicated to Thomas Thynne, Viscount Weymouth, grandmaster of the Freemasons. According to the *Advertisement*, these were "some lectures, privately read to a set of gentlemen," with the mechanical drawings done by Isaac Ware, an architect. Desaguliers is thanked profusely. For the steam engine, see pp. 67–70.

24. Martin Clare, *Youth's Introduction to Trade and Business*, 5th ed. (London, 1740), pp. 97, 109–110.

25. John S. Allen, "Thomas Newcomen (1663/4–1729) and his Family," *Transactions of the Newcomen Society*, vol. 51 (1979–1980), p. 19.

26. W. Brockland and F. Kenworthy, eds., *The Diary of Richard Kay, 1716–51 of Baldingstone, Near Bury: A Lancashire Doctor* (Manchester: Chetham Society, 1968), vol. 16, 3rd ser., p. 63, entry for June 22, 1743.

27. Ibid., p. 26, entry for February 24, 1738/9.

28. "Observations and Memorandums," fols. 1–2 and 8, Chetham's Library, Manchester. These are sometimes cited as being by John Rotheram, but the writer makes it clear that he is a listener; cf. Musson and Robinson, *Science and Technology*, 103n.

29. "Observations and Memorandums," fols. 20–21. In fol. 24 a mathematical illustration of the moon's effect on the tides was given, but the listener records "but whether or no my ignorance as to mathematicks may not be the reason that I don't form a just notion of this calculation I'll rather allow than Dispute." In fol. 35 we are told that "I have in this as well as in all the observations I made on Mr.

Rotherham's Lectures omitted inserting the mathematical experiments and only deduced such rules from them as were servicable [sic] to conducting me through the nature of others."

30. Ibid., fols. 12–14, 57.

31. University Library, Cambridge, MSS ADD. 6301, fol. 6.

32. For notes taken on the lectures of Gervase Holmes (M.A. 1722) at Emmanuel College, see University Library, Cambridge, MSS ADD. 5047.

33. N. Alcock, *Some Memoirs of the Life of Dr. Nathan Alcock, lately deceased* (London, 1780), pp. 7–9.

34. See the excellent treatment of this literature in G. S. Rousseau, "Science Books and Their Readers in the Eighteenth Century," in Isabel Rivers, ed., *Books and Their Readers in Eighteenth-Century England* (Leicester: Leicester University Press, 1982), pp. 197–255. This brings up the whole genre of subscription literature; see Peter Wallis, "British Philomaths—Mid-Eighteenth Century and Earlier," *Centaurus*, vol. 18 (1973), pp. 301–314; and F. J. G. Robinson and P. J. Wallis, "A preliminary guide to book subscription lists: Part 1 pre-1901," *History of Education Society Bulletin*, no. 9 (1972), pp. 23–54.

35. Madame de Cambon [Marie Gertruide de Cambon-van Werken], *Young Grandison: A Series of Letters from Young Persons to their Friends*, translated from the Dutch, 2 vols. (London, 1790), p. 78.

36. Ibid., p. 32.

37. There is no general study of these societies. See, however, Roy Porter, "The Enlightenment in England," in R. Porter and M. Teich, eds., *The Enlightenment in National Context* (Cambridge: Cambridge University Press, 1981), pp. 1–18; R. B. Schofield, *The Lunar Society of Birmingham* (Oxford: Oxford University Press, 1963); Guy Kitteringham, "Science in Provincial Society: The Case of Liverpool in the Early Nineteenth Century," *Annals of Science*, vol. 39 (1982), pp. 329–334; Roger L. Emerson, "The Philosophical Society of Edinburgh, 1737–47," *British Journal for the History of Science*, vol. 12 (1979), pp. 154–191; J. H. Thornton, "The Northampton Philosophical Society, 1743," lecture given to the Northamptonshire Natural History Society and available from the author, to whom I am grateful for assistance; and by way of comparison, James Meenan and Desmond Clarke, eds., *The Royal Dublin Society, 1731–1981* (Dublin: Gill and Macmillan, 1981). Studies that are somewhat related to this topic and should be consulted are F. W. Gibbs, "Robert Dossie (1717–1777) and the Society of Arts," *Annals of Science*, vol. 7 (1951), pp. 149–172; David D. McElroy, *Scotland's Age of Improvement: A Survey of 18th Century Literary Clubs and Societies* (Seattle: Washington State University Press, 1969); Kenneth Hudson, *Patriotism with Profit: British Agricultural Societies in the Eighteenth and Nineteenth Centuries* (London: Hugh Evelyn, 1972), especially pp. 18–23; R. S. Watson, *A History of the Literary and Philosophical Society of Newcas-*

tle-upon-Tyne (Newcastle-upon-Tyne, 1897); and D. G. C. Allan, *William Shipley, Founder of the Royal Society of Arts* (London: Scolar Press, 1979), especially pp. 30–39.

38. For the introduction to the writings of one such historian, Augustin Cochin, see François Furet, *Interpreting the French Revolution* (Cambridge: Cambridge University Press, 1981), Chapter 3; of Cochin's various writings, I have found *La Revolution et la libre pensée* (Paris, 1924) to be the most useful, and some of his ideas are applied here.

39. McElroy, *Scotland's Age of Improvement*, pp. 14–15.

40. Peter Borsay, "The English Urban Renaissance: The Development of Provincial Urban Culture, c. 1680–1760," *Social History*, vol. 5 (May 1977), p. 593.

41. There is no modern account of the eighteenth-century society. See Charles R. Weld, *A History of the Royal Society*, 2 vols. (London, 1848); T. E. Allibone, *The Royal Society and Its Dining Clubs* (Oxford: Oxford University Press, 1976); H. Lyons, *The Royal Society, 1660–1940* (Cambridge: Cambridge University Press, 1944); Charles Lyte, *Sir Joseph Banks, Eighteenth Century Explorer, Botanist and Entrepreneur* (London: David and Charles, 1980), which is largely useless; and H. Hartley, ed., *The Royal Society* (London: Royal Society of London, 1960). One approach to the society by a critic can be discovered in G. S. Rousseau, ed., *The Letters and Papers of Sir John Hill 1714–1775* (New York: AMS Press, 1982); cf. L. Trengove, "Chemistry at the Royal Society of London in the 18th Century," *Annals of Science*, vol. 19 (1963), pp. 183–237.

42. Royal Society, Certificates, vol. 1, 1731–1750; vol. 2, 1751–1766 (approximately 1,000 folios).

43. Ibid., vol. 1, fols. 21, 50, 62, 66, 163.

44. Ibid., fol. 85.

45. Ibid., fols. 95, 118, 139, 167. Fol. 248 relates to a Robert James, "Doctor of Physick," who was rejected; in general, surgeons did not fare well in the voting process. In an essay published after this chapter was completed, M. Crosland draws similar conclusions about the criteria for membership: M. Crosland, "Explicit Qualifications as a Criterion for Membership of the Royal Society," *Notes and Records of the Royal Society*, vol. 37 (1983), pp. 167–187.

46. Royal Society, Certificates, vol. 2, fol. 467.

47. Anon. [P. H. Maty et al.], *An History of the Instances of Exclusion from the Royal Society* (London, 1784), pp. 3 and 10. For further evidence of the power wielded by Banks, see a letter dated October 13, 1805, in Bristol Record Office, MS 8030(1-10), on making someone a fellow: "It is in vain to make the attempt unless Sir Joseph be satisfied."

48. Royal Society, MS 702; see also K. Ochs, "The Failed Revolution in Applied Science: The Study of Industry by Members of the Royal Society of London, 1660–88" (Ph.D. diss., University of Toronto, 1981).

49. Royal Society, MS. L+P. 3, fol. 403.

50. This discussion of the Spalding Gentlemen's Society is deeply in-

debted to Raymond James Evans, "The Diffusion of Science: The Geographical Transmission of Natural Philosophy in the English Provinces, 1660–1760" (Ph.D. diss., University Library, Cambridge, D. Phil., #12208, 1982). Cited with the kind permission of the author.

51. Ibid., pp. 172–175.
52. The manuscript minutes are currently housed at the Spalding Gentlemen's Society, 9 Broad Street, Spalding. Maurice Johnson's MSS, Drawer 1, are particularly interesting on the scientific content of the society's meetings and his contacts with Cromwell Mortimer and the Royal Society.
53. On Stukeley, see Stuart Piggott, *William Stukeley, an Eighteenth Century Antiquary* (Oxford: Clarendon Press, 1950); see also his MS commonplace book at the Spalding society, entry for August 15, 1730, on Daniel's prophecies; and for an unadulterated sampling of his mystical tendencies, see Stukeley MSS, Library of the Grand Lodge, London, MS 1130, "The Creation," fol. 171, on experiments done by Newton "by cutting the heart of an Eel into three pieces" (but that is incidential to the text); and his many manuscripts there on Solomon's Temple. On Grundy, see Royal Society, LBC. 25, fol. 138, Grundy to Senex, 1739 (Senex was also a Freemason); and minutes of the Spalding society and the Institution of Civil Engineers; London, John Grundy MSS, "Surveys, levels, etc.," vol. 2, 1740.
54. Evans, "The Diffusion of Science," pp. 288–290; A. R. Hewitt, "A Lincolnshire Notable and an Old Lodge at Spalding," *Ars Quatuor Coronatorum*, vol. 83 (1970), pp. 96–101.
55. Evans, "The Diffusion of Science," p. 243.
56. G. J. Hollister-Short, "The Introduction of the Newcomen Engine into Europe," *Transactions of the Newcomen Society*, vol. 48 (1976–1977), pp. 11–22.
57. Evans, "The Diffusion of Science," p. 269.
58. Ibid., p. 278.
59. See Eric Robinson, "The Profession of Civil Engineer in the Eighteenth Century: A Portrait of Thomas Yeoman, F.R.S. (1704?–1781), *Annals of Science*, vol. 18, no. 4 (1962), pp. 195–216; A. W. Skempton, "Early Members of the Smeatonian Society of Civil Engineers," *Transactions of the Newcomen Society*, vols. 44–45, (1971–1973), pp. 23–47. There are some Yeoman letters at the Institution of Civil Engineers, London, in John Smeaton's Letter Book.
60. See Philip Doddridge, *A Course of Lectures on the Principal Subjects in Pneumatology, Ethics and Divinity* (London, 1763; published posthumously), especially pp. 43–53 for an attack on Toland and materialism drawn straight out of Clarke's Boyle lectures. Cf. Malcolm Deacon, *Philip Doddridge of Northampton, 1702–51* (Northampton: Northamptonshire Libraries, 1980).
61. Spalding Gentlemen's Society, a loose flier entitled "Answer to the Question the Sixth in the *Ladies' Diary*, for the Year MDCCLI" from William Burwell, a schoolmaster in Norfolk. Cf. Gerald D. Meyer,

The Scientific Lady in England, 1650–1760 (Berkeley: University of California, 1955), *English Studies,* no. 12.

62. H. W. Dickinson, *A Short History of the Steam Engine,* with a new introduction by A. E. Musson (London: Cass, 1963), Chapter 4, especially p. 62.

63. Terry S. Reynolds, "Scientific Influences on Technology: The Case of the Overshot Waterwheel, 1752–54," *Technology and Culture,* vol. 20 (1979), p. 285; cf. A. E. Skempton, ed., *John Smeaton, F.R.S.* (London: Telford, 1981), where none of the contributors discusses Smeaton's theoretical interests or religious beliefs or political values.

64. British Library, MSS ADD. 30,094, fol. 10, Smeaton to Wilson, July 20, 1745. Cf. Royal Society MSS, Treasurer's Book, 1746–1766, where the sum of £23, is commonplace because of the practice of advancing £21 "in lieu of giving Bond," plus £2 admission fee. See entry for 1753 where Smeaton paid the new admission fee of £5.50 plus £21.

65. John Smeaton, *Diary of his Journey to the Low Countries 1755, from the original manuscript in the Library of Trinity House, London* (Leamington Spa, U.K.: The Newcomen Society, 1938), p. 14; and on providence, Skempton, *Smeaton,* p. 22.

66. Smeaton, *Diary,* pp. 5–6, 40; and see the Institution of Civil Engineers, London, John Smeaton Letter Book, fols. 47, 50, 58.

67. British Library, MSS ADD. 30,094, fol. 46, Smeaton to Wilson, July 23, 1747.

68. Ibid., fols. 42, 46, 69, 74, 112; and see Denis Smith, "The Professional Correspondence of John Smeaton: An Eighteenth-Century Consulting Engineering Practice," *Transactions of the Newcomen Society,* vol. 46–47 (1973–1976), p. 181.

69. Institution of Civil Engineers, John Smeaton Letter Book, fols. 50, 132; cf. Smith, "Correspondence of John Smeaton," pp. 179–189; and A. W. Skempton, "Early Members of the Smeatonian Society of Civil Engineers," *Transactions of the Newcomen Society,* vol. 44–45 (1971–1973), pp. 23–47.

70. The Institution of Civil Engineers, John Smeaton Letter Book, fol. 39. The letter is clearly by Smeaton but he signs it "Agricola"; it is to a Mr. Ledger. See also Skempton, *Smeaton,* pp. 4, 54, 249; and for Smeaton's own concern for the cost of labor, see Institution of Civil Engineers, Smeaton Letter Book, fol. 32. Cf. Royal Society, L+P VII, no. 11, fols. 4–5, for Smeaton's interest in water and steam engines as replacements for horsepower.

71. Only the Rev. Timothy Neve, founder of the Peterborough society, had little trust in "the religious capacity of the Court"; but he dined with Walpole at Houghton Hall and liked it. See Spalding Gentlemen's Society, Maurice Johnson MS, Drawer 1, no. 14, letter from the late 1720s.

72. E. Fearn, "The Derbyshire Reform Societies, 1791–93," *Derbyshire Archaeological Journal,* vol. 83, (1968), pp. 48–55.

73. See Henry Redhead Yorke, *Reason urged against precedent* (London,

1793), for a full statement of radical sentiments commonplace in these circles. Yorke eventually grew disillusioned with the Derby corresponding society and moved on to the more radical one at Sheffield. On Darwin, see Brian Easlea, *Science and Sexual Oppression* (London: Weidenfeld and Nicolson, 1981), pp. 94–99.

74. See the printed *Rules and Catalogue of the Library Belonging to the Derby Philosophical Society* (Derby, 1815). Even more interesting is the manuscript list of acquisitions numbered in order of acquisition; and the list of borrowing shows that members also attempted to read, not just buy, the society's books. Derby Borough Library, MS BA 106.

75. Joyce Appleby, "What Is Still American in the Political Philosophy of Thomas Jefferson," *William and Mary Quarterly*, 3rd ser., vol. 39, no. 2 (1982), pp. 308–309; cf. Appleby, *Capitalism and a New Social Order* (New York: New York University Press, 1982).

76. Desmond King-Hele, *Doctor of Revolution: The Life and Genius of Erasmus Darwin* (London: Faber and Faber, 1977), pp. 60–62. Jefferson described Small as having given him "my first views of the expansion of science and of the system of things in which we are placed."

77. *Address to the Philosophical Society . . . July 18, 1784,* by Doctor Darwin, president, found in *Rules and Catalogue,* pp. ix–xiv.

78. Eric Robinson, "The Derby Philosophical Society," *Annals of Science,* vol. 9 (1953), p. 360. On Pigott, see James O. Manton, *Early Freemasonry in Derbyshire* (Manchester: Marsden, 1913), p. 21. See also D. King-Hele, ed., *The Letters of Erasmus Darwin* (Cambridge: Cambridge University Press, 1981), p. 128.

79. M. R. Austin, "Religion and Society in Derbyshire During the Industrial Revolution," *Derbyshire Archaeological Journal,* vol. 93, (1973), pp. 75–89.

80. *William Strutt—Memoir,* typescript in Derby Local Library, no. 3542, p. 27.

81. Philo-Filmer, *Encomiastic advice to the Acute and Ingenious Personage who parodied the Address from the Derby Societies to the Rt. Hon. Charles James Fox* (London, 1793), p. 4; on p. 13: "At the epithets Leveller, and Republican, they smile." This is a satire by someone actually in favor of the societies, probably William Ward. Cf. *The Derby Address: At a Meeting of the Society for Political Information held at the Talbot Inn, in Derby, July 16, 1792,* "To the Friends of Free Enquiry, and the General Good."

82. M. C. Egerton, "William Strutt and the Application of Convection to the Heating of Buildings," *Annals of Science,* vol. 24 (1968), pp. 73–88. Cf. *William Strutt—Memoir,* pp. 17–20; and Charles Sylvester, *The Philosophy of Domestic Economy* (Nottingham, 1819), dedicated to Strutt.

83. *William Strutt—Memoir,* p. 60. The poet Thomas Moore found them "true Jacobins" in 1814; see p. 10 for copy of a letter by Elizabeth Evans, Strutt's sister, on Godwin; and a letter of condolence from Elizabeth Darwin to William Strutt in 1804 (p. 47) never mentions

God or providence. I wish to thank Margaret Hunt for assistance with this archive and also the excellent staff of the Derby Local Library, Karen Smith and Sylvia Gown.

84. *Letters of Josiah Wedgwood 1762–1772* (London, 1903), pp. 54–55.
85. Ibid., p. 105.
86. Ibid., p. 203.
87. Ibid., p. 217.
88. Eric Robinson and A. E. Musson, *James Watt and the Steam Revolution* (London: Adams and Dart, 1969), pp. 204–205, which prints a manuscript entitled "Points necessary to be known by a steam engineer," 1796.
89. E. P. Thompson, *The Making of the English Working Class* (London: Gollancz, 1963), pp. 738–740; cf. S. Shapin and B. Barnes, "Science, Nature and Control: Interpreting Mechanics' Institutes," *Social Studies in Science,* vol. 7, pp. 31–74. On radicalism see Isaac Kramnick, "Eighteenth Century Science and Radical Social Theory: The Case of Joseph Priestley's Scientific Liberalism," *Journal of British Studies,* vol. 25, no. 1 (1986), pp. 1–30.

CHAPTER 6

Scientific Education and Industrialization in Continental Europe

One of the most persistent myths about the Industrial Revolution in the Western world assumes that "most of the inventions in the early stages of the industrial revolution were not made by scientists, but by people with little or no education, who experimented to find new and better solutions to urgent problems."[1] Given all that has been discussed in the previous chapters about the process by which scientific knowledge became integrated into European elite culture by the early eighteenth century, that statement should now appear to be naive; it is also hopelessly misleading. When offered as a model for third world development today—as it frequently is—that prescription for success dooms its followers to failure. The statement is historically false because it presumes a distinction between the "scientist" and all others that simply did not exist in the late eighteenth century when industrialization began, first in England and then in Belgium and Switzerland.

Of the hundred or more leading English scientists of the century, for example, nearly half would have to be classified as "devotees" (to avoid the anachronistic term "amateurs"), and of that hundred, 45 percent made their living as doctors, technicians, or churchmen.[2] In addition the falsity of the statement arises from the naive assumption that somehow human beings are born to think "mechanically," in the post–Scientific Revolution meaning of the term. In consequence "the people of little or no education," who supposedly made the Industrial Revolution, are presumed to have understood mechanical principles of nature, to have thought about steam and water as mechanically controllable matter without having been in some manner educated to that assumption.

Yet everything we know about European intellectual history, from the crisis identified by Hazard onward, attests to the all-pervasive nature of scientific interest and knowledge among the literate classes in almost every western European country. That knowledge grew exponentially, if somewhat unevenly, from the 1720s to the 1790s, and this chapter seeks to document and analyze its growth in Continental Europe.

One way of documenting the impact of scientific thinking would be to catalogue the complex ways in which science was used by the intellectual leaders, or philosophes, of the European Enlightenment. Many fine books have traced with insight the exact relationship between science and the Enlightenment.[3] They have argued persuasively that in the eighteenth century scientific thought provided a model for social thought, that the European philosophes believed that the successful use of human reason in the natural world made possible its successful application to the social order. But it is also possible to approach eighteenth-century culture from another perspective, one that may be self-consciously described as somewhat "Whiggish"—by which is meant that the historian asks of the past questions that presume an outcome incapable of being known, an outcome perhaps even undesired, by the historical actors themselves. In general that is not a particularly useful way to approach any epoch. Yet it does seem reasonable to approach eighteenth-century scientific culture in Continental Europe with the realization that by the 1790s industrialization had begun there in certain places and not in others. It seems reasonable in turn to inquire what, if anything, scientific knowledge had to do with that historical transformation, bearing in mind that in certain parts of western Europe, where essentially feudal and agrarian society predominated, no amount of scientific knowledge would have propelled those areas in an industrial direction.

To use another approach to this matter of science and its role in the industrialization of Continental Europe, can we understand why the Industrial Revolution did not occur in those few places where other conditions imply that it "should" have occurred in the late eighteenth century? To economic and social historians, long schooled to believe that economic and social factors alone must account for the making of industrial Europe, that question would seem largely irrelevant to a process that began in England because of material conditions and was "delayed" on the Continent for comparable material, and non-cultural, reasons. But the historical narrative of industrialization is far more complex than that.

When we compare the penetration of scientific and mechanical knowledge in Continental Europe with what was occurring in England during the same period (see Chapter 5), the question we now pose yields some rather startling insights. Scientific knowledge (i.e., cultural factors) cannot *explain* the Industrial Revolution in one place or another in the period from 1750 to 1820; but then neither can purely economic or social factors. Real people, not "factors," made the Industrial Revolution on both sides of the Channel, and in order to mechanize, those people had to be able to think mechanically. The evidence, especially from The Netherlands but also from France—the two countries where industrialization might have occurred prior to 1800—strongly suggests that many of the very men who had access to capital, cheap labor, water, and even steam power could not have industrialized had they wanted to: they simply could not have understood the mechanical principles necessary to implement a sophisticated assault on the hand manufacturing process. By and large, as we shall see in the concluding chapters, their British counterparts did possess that knowledge, and they put it to effective use from the 1760s onward. From the perspective of scientific knowledge and education of a mechanical sort, the English elite was at least a generation ahead of its European counterpart.[4] That generation of entrepreneurs who flourished from 1760 to 1800 proved critically important in providing Britain with an industrial head start, nothing more, nothing less.

A balanced account of the historical conditions within which industrialization might occur must add scientific knowledge of a specifically mechanical sort to a mélange that includes a source of natural power, cheap and exploitable labor, a market for the consumption of goods, and so forth—all the various material factors that have just been described and that we will see at work in the British Industrial Revolution (see Chapter 7). There we will see some of the actual uses to which mechanical knowledge was put in the critically important early stage of that industrial process. Here the integration of scientific knowledge in Continental Europe is being approached with that "British model" self-consciously in mind. Was that mechanical knowledge as widespread in those areas where the material conditions for industrialization may be said to have existed? And given what we now know about the subtle but vital role played by central government in the British process—by the parliamentary system that evolved out of the English Revolution—we must add political factors to those "material" conditions with which we must reckon. In many parts of northern and western Europe it was not enough for some segments of the literate elite to possess the knowledge and will to exploit man and nature, it was also necessary to have a central government responsive to their interests.

SCIENCE AND THE "DECLINE" OF THE DUTCH REPUBLIC

The evidence from very late in the eighteenth century in both France and The Netherlands suggests that scientific reformers of an industrial sort saw the process of political revolution that engulfed both those societies, beginning in 1787 in Amsterdam and 1789 in Paris, as directly beneficial to promoting the knowledge they championed and sought to inculcate. Indeed it was some of those very reformers who were the first to notice and describe the gap that existed between Britain and their own countries in the area of applied scientific knowledge. Dutch reformers, in particular, recognized the gap and self-consciously sought to close it, to imitate British industrialists such as Josiah Wedgwood, who had used their knowledge and their capital to improve—we would say "to industrialize"—the manufacturing process. The insights of those reformers, coupled with other evidence, adds another dimension to one of the thorniest problems frequently discussed by historians of Western industrialization—namely, why did the Dutch Republic fail to industrialize in the late eighteenth century?

It should be evident from the discussion in previous chapters (see pp. 52–53) that The Netherlands had been one of the most scientifically advanced areas in seventeenth-century Europe. The Dutch scientists Beeckman and Huygens, among others, ranked with the leading mechanists of their respective generations, while the Dutch universities responded first to Cartesianism and then to Newtonianism in advance of other Continental centers of higher learning. While this may not be as striking in the case of Cartesianism, because its penetration can also be observed in the Spanish Netherlands by the 1670s, it is unequivocally striking for the rapidity with which Newtonianism was accepted in the Dutch Republic. By comparison, the leading Belgian university at Leuven (Louvain—first under Spanish then under Austrian domination) enthroned Descartes in the 1670s, only to leave the statue untarnished, never mind untoppled, until well into the eighteenth century.[5] Similarly, Dutch lens grinding and superior optical work created the milieu wherein Anton Leeuwenhoek invented the microscope, and Leiden excelled in the early modern period as a center for medical education. No Continental country possessed a freer press or easier access to scientific treatises.

Yet by the 1740s Dutch eminence in scientific and technical knowledge—so conspicuous in the seventeenth century—had given way to an increasing anxiety about its place in the republic of learning, an anxiety founded, it would seem, on concrete factors at least as far as mechanical knowledge is concerned. High on the list stood the condition of the Dutch economy at midcentury, which now conspicuously lagged against the performances of foreign competitors as well as against its own late seventeenth-century performance. Yet declining prosperity, if confined to certain economic and social groups, need not inhibit the onset of industrialization. Indeed as much as any area in western Europe and perhaps more so, the Dutch Republic, particularly its manufacturing centers around Haarlem, Leiden, and its coal-laden southern province, Limburg, have been assessed as ripe in many ways for industrialization. Vast amounts of capital drawn from the commercial revolution of the sixteenth and seventeenth centuries could be found among elite as well as mercantile families, while the poverty and hence exploitability of the rural and urban masses increased in the course of the eighteenth century. In addition strong evidence exists to suggest that where that commercial elite needed advanced technology, in particular instruments for navigation, it managed to get them.

In 1778 a Dutch newspaper defined the country as a nation of "rentiers and beggars," and however exaggerated, the phrase suggests a great deal.[6] The term "rentier" denotes those who lived off their rents or investments, through the profits of commercial transactions, rather than those who generated capital through productive entrepreneurial activity. Indeed one of the earliest negative uses in Dutch of the word "capitalist" occurred in this period, and it defined such people as "rentiers."[7] Beggary was also common, especially by mid-century when the clothing industry declined as a result of foreign competition. In other words, an impoverished class existed in parts of the republic that could have been proletarianized, as happened in Britain and the southern Netherlands (i.e., Belgium). And there was certainly no shortage of capital in what had once been the richest nation, per capita, in all of Europe. What appears to have been missing in significant numbers were entrepreneuring capitalists interested in the industrial process. Their absence has led some historians to argue for psychological factors as the key to the Dutch "decline," relative to its more competitive neighbors.

That psychological argument can be countered by the traditional objection that capitalist and exploitable worker do not an industrial revolution make. Iron and coal, both lacking in significant quantities in the Dutch Republic, particularly in the province of Holland, were equally vital. While the importance of those raw materials should not be minimized, it must be noted that their place in the British Industrial Revolution would have been far less important had it not been for the engineering feats that permitted their efficient exploitation: the canal building of the 1770s to 1790s, the implementation of steam power,[8] and more important, the harnessing of water power through mechanical devices. Indeed the most current thinking among some economic historians assumes that no single factor or narrowly defined group of factors can adequately account for industrialization. When local studies are employed we find a mélange of factors used, or not used, by entrepreneurs; even more fascinating, we find the presence of the necessary material conditions and no entrepreneurs willing or able to exploit them.

To turn then to scientific and mechanical knowledge as one highly underrated element in the process of industrialization seems especially appropriate.[9] Eventually, and quite rapidly, mechanical knowledge spread everywhere in European culture, only it did so more slowly in certain parts of western Europe than it did in Britain. That time lag demands our attention especially because the earliest Dutch Newtonians—and they were the first European Newtonians—sought to imitate their British counterparts, and in the case of William Jacob s'Gravesande even excelled as popularizers and promoters of applied mechanics.

In Continental Europe those early Dutch scientists—such as Boerhaave, s'Gravesande, and Petrus van Musschenbroek (1692–1761)—learned this revolutionary modification of the mechanical philosophy directly from the master himself or from his immediate associates and followers, such as Samuel Clarke or Archibald Pitcairne, who was professor of medicine at Leiden in 1693. They sought in turn to displace Cartesianism once and for all from the Dutch university curriculum. As Musschenbroek put it in a letter to the now aged Newton (d. 1727):

> Being an admirer of your wisdom and philosophical teaching, of which I had experience while in Britain in familiar conversation with yourself, I thought it no error to follow in your footsteps (though far behind), in embracing and propagating the Newtonian philosophy. I began to do so in two universities where the triflings of Cartesianism flourished, and met with success, so that there is hope that the Newtonian philosophy will be seen as true in the greater part of Holland, with praise of yourself. It would flourish even more but for the resistance of certain prejudiced and casuistical theologians. I have prepared a compendium for beginners with which, if it does not displease you greatly, I shall be well satisfied. I shall always endeavour to serve the wisest man to whom this earth has yet given birth.[10]

Musschenbroek had been in London in 1719, and he had proceeded on his return to The Netherlands to teach Newton's system at Duisberg and Utrecht.[11] His fellow Newtonian, s'Gravesande, like Boerhaave before him, embraced a similar project at Leiden, having also learned the new mechanical philosophy from its master. In 1718 s'Gravesande wrote to Newton:

> I begin to hope that the way of philosophizing that one finds in this book will be more and more followed in this country, at least I flatter myself that I have had some success in giving a taste of your philosophy in this university; as I talk to people who have made very little progress in mathematics I have been obliged to have several machines constructed to convey the force of propositions whose

demonstrations they had not understood. By experiment I give a direct proof of the nature of compounded motions, oblique forces and the principal propositions respecting central forces.[12]

Like his British counterparts, s'Gravesande had encountered mathematical ignorance among his countrymen and responded by illustrating the Newtonian universe with mechanical devices. In this he was similar to his intimate associate Desaguliers, who also gave his mechanical lectures in the Dutch Republic (probably in French), where they were then translated into Dutch and published.[13] s'Gravesande also shared Desaguliers's enthusiasm for the industrial application of mechanical devices and his interest in the early steam engine.[14] Indeed part of s'Gravesande's obligations as professor of natural philosophy at Leiden—a position secured for him through Newton's intervention—included the surveying and improvement of water transportation in the republic.[15]

Not least, s'Gravesande belonged to a circle of publishers and journalists, many of them French Huguenot refugees, who were singularly important in transmitting the Newtonian philosophy through their French language journals. They in turn were among the few citizens or residents of the Dutch Republic to be made Fellows of the Royal Society.[16] s'Gravesande's circle in Leiden and The Hague may now be counted as the first anywhere in Continental Europe to accept Newtonian science wholeheartedly and to promote it aggressively. In distant outposts of the Dutch empire such as Surinam, the propagandizing efforts of this circle, constituted as a private literary society, were felt as early as 1723.[17] And most important, those propagandizing efforts were in French, the language of most literate elites in eighteenth-century Europe, as well as in Dutch. s'Gravesande's circle is also interesting for another reason: many of its members, and possibly even the professor himself, became Freemasons. That secular fraternity played an important role in the eighteenth century as the promoter of scientific education in the Low Countries. Indeed we may take it as a working hypothesis—one upon which further research will someday have to be done—that in eighteenth-century Europe Freemasons played a role in relation to scientific education analogous to that of progressive Calvinists in the seventeenth century. In disproportionately large numbers

Freemasons promoted the new science by organizing lectures and philosophical societies for scientific devotees and laity like themselves. In so doing they exercised a role as progressive improvers, as the concrete promoters of the highest of Enlightenment ideals.

Also out of s'Gravesande's Leiden classroom came the next generation of Dutch Newtonians, who took this mechanically explicated science to other Dutch colleges and universities, to Franeker and Harderwijk, for example, as well as to Amsterdam. The public lectures given there in 1718 were by Fahrenheit (made famous by his system for measuring heat), and he had worked closely in mechanics and the use of mechanical devices with s'Gravesande.[18] And not least, s'Gravesande's influence lasted to the end of the century in the main Dutch scientific society at Haarlem and in the scientific thought of the Newtonian and revolutionary reformer, J. H. van Swinden (see pp. 196–197). Voltaire admitted that he learned a great deal from s'Gravesande's published explication of Newton's system, as did the most important French public lecturer of the first half of the century, the abbé Nollet[19] (see pp. 200–202).

Yet despite this significant head start in scientific education, more precisely in mechanically oriented explications of Newton's system, the Dutch republic by mid-century evinced no widespread program of popular scientific education aimed at elite audiences by comparison to efforts visible across the Channel in the second half of the century.[20] Partly as a result of opposition from the University of Leiden, Dutch scientific societies began to be formed only after 1752, when De Hollandsche Maatschappij der Wetenschappen (the Holland Society for Science) was founded in Haarlem.[21]

The Holland Society, unlike almost all other scientific academies on the Continent, was a private body without an official relationship to the government, and certainly without one to the king—the Dutch stadholderate could hardly be described as an absolute monarchy similar to those found, for example, in France, Spain, Prussia, or Russia. The Haarlem society was supported, as was the Royal Society of London, by the dues of its members, and as such may be expected to reflect their immediate interests more closely than did other academies officially licensed by the crown. One of the signal characteristics of those scientific societies in absolutist countries was the remarkable accuracy with which they reflected the dominance of the old order, more precisely of the nobility, as well as the material interests of the government. That is not to say that they did less science, or even less "pure" science. Indeed the level of original scientific inquiry was probably higher at the Academy of Science in Paris than it was in Haarlem or London, but that is not our concern. Where the lay elite could control the pursuit of scientific knowledge it basically got the kind of science that it wanted.

A survey of the Dutch Society's proceedings during the first few decades of its existence reflects its interests and, incidentally, reveals that its largely clerical, commercial, aristocratic, legal, and medical members favored certain kinds of scientific inquiry over others. Christian natural religion, or physico-theology, was commonplace in their discussions, as were sophisticated astronomy and the latest medical problems. Some applied mechanics were explicated in the manner of s'Gravesande or Desaguliers, but this was a minor aspect of the society's transactions. There was also a predictable interest in canals and dikes, as well as most notably in navigation, although little mention is made of foreign innovations in hydraulics and hydrostatics. Like most European scientific societies or academies, the society posed annual questions for which prizes were given; yet significantly only in 1787 did it turn its attention to the question of the relationship between industry and commerce. In that year of revolution, however, no essay answers were submitted. In those areas where the members excelled, their scientific knowledge was extensive. It is only from the Whiggish perspective adopted here that emphasis is laid on the absence of practical and applied mechanics.[22] That was not their interest. Only because we are trying to answer larger questions in the history of European development are we justified in commenting on the relative myopia of the leading Dutch scientific society.

That myopia can also be observed in other areas of Dutch scientific education in the eighteenth century. Aside from those academies where students of s'Gravesande can be found, there is evidence especially from the middle decades of the century to suggest a notable failure to keep abreast of the latest knowledge in the physical sciences, particularly of an applied and mechanical kind. The exception to this general pattern appears to have been among the Mennonites and Anabaptists in the Republic. The former group had requested Fahrenheit's Amsterdam lectures in 1718, and they had also institutionalized the teaching of science in their seminary after 1740.[23] Among these dissenters from rigid Calvinism the tradition of physico-theology imported from the early English Newtonians seems to have been instrumental in reconciling their highly individualistic form of piety with the new science.[24]

By contrast, in the library of the academy of Harderwijk, where the new science is very much in evidence during the second half of the seventeenth century, emphasis in the eighteenth century appears to have been legal, medical, and theological rather than scientific or mechanical, with the notable exception of works by s'Gravesande and Musschenbroek.[25] Only very late in the eighteenth century do we begin to see evidence in the province of Gelderland for the existence of public scientific lecturing intended for commerce, trade, and industry, and these were associated predictably with the local scientific academy.[26] A similar lack of interest in science also plagued the academy at Deventer, and progressive parents in turn sent their children elsewhere on the Continent or to Amsterdam, where by the 1760s public agitation for reform in scientific education began in earnest.[27] In Deventer the local Calvinist clergy appear to have been particularly powerful at the academy and to have maintained a curriculum that had been innovative in the seventeenth century but was anachronistic by the mid-eighteenth. While Calvinism in the seventeenth century may have produced scientific rationalists such as Beeckman, by the eighteenth its orthodox clergy had grown fearful of heresy among the laity, and the power of Calvinist orthodoxy in popular culture produced widespread public opposition to aspects of the new science, for example, smallpox inoculation.[28]

The absence of government intervention may also be significant in trying to explain the relative failure of Dutch Newtonians and their English allies such as Desaguliers to spread the Newtonian gospel among the mercantile and to encourage the mechanically industrious. Desaguliers was one of the earliest promoters of the Newcomen steam engine, and in this s'Gravesande appears to have assisted. Indeed he was consulted in 1721 by engineers attempting to install an atmosphere engine in Germany for the landgrave of Hesse. Likewise Belgian mine owners as early as the 1720s employed an engineer trained in part by s'Gravesande, and in the area around Liège various attempts were made in the course of the century to harness the power of the new engines.

Essential to these efforts was the encouragement offered by governments as well as the presence of engineers trained in the Newtonian tradition of applied mechanics so commonplace in the lectures of the itinerant propagators of the new science.[29] In the Dutch Republic, unlike the Austrian Netherlands (i.e., Belgium), that kind of central authority was simply not part of the constitutional structure, and in its absence private efforts at technological improvements based on the new science—such as we find in the eighteenth century in Britain—would have had to be very widespread.[30]

In 1751 when an entrepreneuring Rotterdam clockmaker

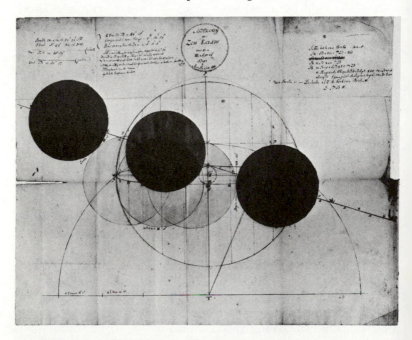

These quite sophisticated notes, taken by a student of J. H. van Swinden in the 1790s, describe an eclipse of the sun. (*Courtesy of the Museum of the University of Amsterdam***)**

sought to install a steam engine he had to go to England to make his inquiries. Although that engine ultimately failed because of a weak and overelaborate mechanical arrangement for connecting its water pumps, the effort did lead to the establishment in 1769 of a scientific society in Rotterdam, which in the 1780s brought Watt's steam engine to The Netherlands.[31]

In the absence of governmental intervention intended to improve technological capacity, which in turn could have encouraged the kind of scientific education for the laity found routinely in Britain, two factors seem critical on the Continent in determining whether or not scientific education would become commonplace and bear industrial fruit. The presence of entrepreneuring scientists with a distinct interest in applied mechanics, although essential, was not sufficient. They required an elite audience willing to pay for their knowledge, and ultimately both would require access to governmental institutions willing and capable of fostering industrial development.

The growth of literacy and printing—two of the most revolutionary forces in the preindustrial period—had created in western Europe, and nowhere more so than in Britain and the Dutch Republic, a marketplace for learning of unprecedented size and importance. In both countries by the late eighteenth century literacy appears to have been possessed by the majority of males —in contrast to France where 20 to 30 percent of the male population may have been literate. The market thus created[32] appears to have been critically important whenever the new science encountered opposition from the entrenched clerical faculties of the colleges and universities. Even where its acceptance had been secured, as was the case in certain of the Dutch universities prior to any others on the Continent, that scientific education touched only a minuscule fraction of the literate population. The world of books, lectures, and scientific societies remained essential, and there elite interest and participation were the keys to success.[33]

In the course of the eighteenth century a multitude of British natural philosophers managed to make a living out of scientific education. Remarkably their Dutch counterparts were thin on the ground prior to the 1760s. The difficulty appears to have lain precisely in the absence of significant interest in scientific and mechanical education among the old commercial elite or the merchant community, particularly outside Amsterdam. The impetus begun by s'Gravesande failed to take root in the next generation and was only revived later in the century by new scientific reformers.

In the late 1740s in Amsterdam, radical critics of the existing political and social order in the Dutch Republic were strident in their condemnation of the corruption and apathy they now attributed to the ruling elite, the so-called *regenten,* who monopolized wealth as well as governmental offices in the towns and cities. They were commenting on and laying blame for what contemporaries were beginning to describe as the century of decline—decline relative, of course, to the brilliant prosperity and inventiveness characteristic of the republic in the seventeenth century, in its so-called *Gouden Eeuw* (golden century).

The concept of the "decline" of The Netherlands may be argued to have been nothing other than the inability of such a small country (population less than 2 million) to compete in an increasingly consumer-oriented Western economy, where its large and more unified rivals, in particular Great Britain and France, possessed access to the necessary markets.[34] Yet with regard to The Netherlands in the eighteenth century the concept of decline as a cultural phenomenon is difficult to dismiss, not least because the charges leveled by contemporary critics appear to be borne out by research in at least one area—namely, the absence of elite interest in scientific education useful for industry. The radicals of the late 1740s pointed precisely at that intellectual lethargy,[35] as well as to the decline in manufacturing. Indeed in 1751 the newly restored stadholder, William IV, set up a commission to study the decline of commercial and industrial activity,[36] but nothing came of the inquiry.

Within the mid-century entourage of the new stadholder, enlightened as it was, scientific learning enjoyed a fashionable place. That enlightened aristocracy at The Hague attended scientific lectures of considerable sophistication,[37] and these give us an opportunity to compare what the lecturer there believed to be of interest with similar lectures routinely given in England (see Chapter 5). At The Hague the new science from Copernicus through Galileo, Kepler, Descartes, Newton, and Leibniz, as well as the electrical experiments of Benjamin Franklin, were explicated as the singular achievement of European civilization.[38] Great care was also taken to praise Descartes, while at the same time the achievements of Newton were carefully recounted, although some of his followers are accused of attempting to reintroduce the occult qualities so carefully avoided by the mechanists of the seventeenth century. Indeed the lecturer is at pains to note that the true physicist is neither Cartesian nor Newtonian and asserts that all true science displays that God is the sole master of nature. Physico-theology is skillfully woven with the assertion that science and mathematics are useful for trade and commerce.[39] Considerable attention is given to chemistry, to Boyle's law, to the phenomena of density and porosity of bodies. The use of the microscope is demonstrated, and contemporary biological theories are also discussed. In the section on dynamics where universal gravitation is explained, among other basic Newtonian principles, recourse is made to fairly sophisticated mathematics.[40] In fact, for our purposes, the most significant part of the lectures consists in the mathematical illustrations freely used and the fact that no actual mechanical devices are introduced. Here we have a superb example of a sophisticated set of scientific lectures, more advanced than what was routinely available to a British lay audience of the period, which offered its listeners what it was believed they could absorb and what interested them. Practical, industrial application seemed to matter little to the aristocrats of The Hague. They did entertain the importance of applications of the new science in metallurgy —for example, in the methods of weighing precious metals[41]— or in the development of mathematical skills useful for trade. In a sense there was far more "pure" and sophisticated contemporary science to be learned at these lectures than at the courses given by Desaguliers and his many British followers. This cosmopolitan audience could be said to have come away from this series better versed in the state of scientific knowledge at the midcentury than their counterparts across the Channel. What they were not taught, however, were the many practical applications in mining and manufacturing to be extracted from simple mechanical devices. The value of mathematics and applied science for commercial transactions had long been recognized by the Dutch elite, indeed even the young and future stadholder was educated in the late 1750s in mathematics directly useful for business.[42] In the transition, however, from commercial to industrial capitalism more was needed than mathematics for trade or astronomy for navigation or physico-theology for the inculcation of piety.

We can see a specifically commercial spirit dominant in the physico-theological literature pervasive in the Dutch Enlightenment, a genre of literature that also, to be sure, appealed to a broad European audience. J. F. Martinet's *Catechism of Nature* (Amsterdam, 1777) went through a multitude of editions in Dutch and then in English and may be said to summarize a commercially focused piety that simply did not see the need to address the question of manufacturing through mechanical applications.[43] In that sensibility, all of nature is arranged hierarchically and intended for human exploitation. The beauty of the heavens complements the order in the animal and vegetative world; commerce and navigation are the keys to prosperity and to the exploitation of nature's riches. "The whole world is a grand storehouse for man"—the gold from Africa (slaves go discreetly unmentioned) as well as the tobacco from America are but examples of its yield. The child or adult instructed by this catechism is told that even if he is not a merchant he (there is no appeal to women) should know what is available and exploitable by means of commerce and navigation. This was a piety that self-consciously harked back to the extremely popular Protestant physico-theology of the early eighteenth century, an intellectual invention that was primarily English in origin but was also reinforced by an independent Dutch literature, much of it in turn translated into English.[44] That earlier physico-theology embraced the world of commerce and sought to Christianize it. Yet it never addressed the possibility of industrialization, nor could it possibly have given expression to social and political discontent. Physico-theology served to ensure political stability and economic progress of a commercial sort. Yet beginning in the late 1750s that stability was threatened by a widespread discontent among reformers and progressives within the Dutch Republic and elsewhere.

By almost any reckoning the Dutch aristocracy of the eighteenth century—however mercantile its origins—was among the most entrenched in Europe. Some two hundred families, many based in Amsterdam, monopolized the senior offices in that city and many others.[45] Their control over both the wealth and the governmental institutions of the Republic complemented the au-

thority of the stadholder's party in The Hague, which was resolutely pro-English. By the 1760s many lesser merchants in Amsterdam were openly hostile to both the *regenten* and the
stadholder. As they explained to an English visitor, "their greatest grievance was to see their country enslaved by their own
countrymen—by the very representatives who were chosen to
protect their liberties and privileges."[46] Although Amsterdam
alone paid the majority of the government's revenue, they maintained, they were still forced to go to The Hague and "dance in
attendance" at the feet of the stadholder and his party. In consequence they were "neither attached to the French nor English at
least no farther than as they connect them with liberty and the
independence of the Republic."[47] And not least "the principal
people in Amsterdam formed an association to shake off every
connection with the rest of the provinces and they did not doubt
but it would soon come to this." Such associations as this English
visitor heard about may have been nothing more than the various
literary and philosophical societies in Amsterdam, where talk
about the problems of "the fatherland" was commonplace by the
1760s.[48] The noticeable interest in science and learning of such
societies stands in marked contrast to the eating and drinking
clubs of the *regenten*,[49] to the display of wealth that seems "to lie
in heaps" in Amsterdam, as another visitor put it. By comparison
to Amsterdam (population 200,000), as that same English visitor
noted, the Protestant city of Edinburgh showed a marked interest
in "the sciences, the arts, and the lore of agriculture."[50] In short
the cries for reform heard in the late 1740s had once again
sounded, only now with the revolutionary consequences that in
the late 1780s proved to be fundamentally threatening to the old
Dutch ruling elite.

These Amsterdam critics of the *ancien régime* advocated that
merchants learn the science and mechanics *(werktuigkunde)* taught
by Desaguliers and the few scientific lecturers, such as Benjamin
Bosma, who had continued the tradition he began.[51] They advocated that a scientific society be established in Amsterdam,
similar to the one in nearby Haarlem, only this one should cater
to merchants and their interests.[52] They advocated a revival of
manufacturing along English lines and asked pointedly, "Why do
the English prosper more than we do in art and science?"[53] And
significantly from the perspective of political history, these na-

tionalistic proponents of mechanical science intended for industry styled themselves as early as the 1770s "patriots," as the leaders of the Dutch Revolution of 1787 would be known.

A similar interest in mechanization for industry can be found within the leading literary-philosophical society of Amsterdam.[54] There lectures were given on French innovations in porcelain manufacturing as well as on the industrial techniques of Wedgwood (see pp. 136, 168), one of the pioneers of the British Industrial Revolution.[55] Indeed throughout the last quarter of the century the non-*regenten* founders of Felix Meritis, as that society was called, demonstrated a marked interest in mechanical techniques intended for industrial application.[56] Similarly one of the other learned societies of the city, Concordia et Libertate, was also reformist and critical of the existing order. It had Benjamin Bosma as a member, and his scientific lectures were among the most practically mechanical to be found in Dutch during the period.[57] His science was very much in the manner of the lectures that were commonplace in Britain; not surprisingly Bosma passionately advocated mechanics as a way of facilitating human labor, and he inveighed against a life of leisure.[58] He also attacked the moribund condition of Dutch science and pointed to Germany, France, and England, where "300 men have excelled in mathematics" while "I can count no more than ten in The Netherlands."[59] The one exception, he claimed, could be found in Amsterdam, where it was possible to find merchants with a genuine interest in the new mechanical science.[60]

Such sentiments about the need for merchants to learn natural philosophy can occasionally be found earlier in the century, complete with an emphasis on applied mechanics for the improvement of windmills, pendulums, and shipping,[61] all for the lessening of the burden of human labor. But in the late eighteenth century the frequency of such pronouncements at public lectures, and particularly in the courses offered at the Amsterdam private societies and at its one advanced college, the Athenaeum, vastly increased. In addition a specifically industrial focus is given to this education, with the Dutch Newtonian scientist J. H. van Swinden (1746–1823) taking the lead with lectures on the steam engine[62] and porcelain making, as well as on the traditional subjects intended for trade and navigation, in particular astronomy.

Indeed at the Athenaeum (which in 1877 became the University of Amsterdam)[63] we can see on its faculty during the last quarter of the eighteenth century a distinct blending of scientific educational reform, with an industrial focus, and political reform. Van Swinden, who became a patriot and an active participator in the Dutch Revolution of the late eighteenth century, exemplifies this tendency to advocate industrial science as part of a larger program aimed at social and political reform. The Amsterdam apothecary Willem van Barneveld (1747–1826), also lectured at the Athenaeum on applied science and in turn became an ardent patriot. Indeed a number of Amsterdam scientists as well as their mechanically minded scientific friends in Rotterdam can be counted as active in the scientific education of the city and then as revolutionary reformers.

We may be reasonably confident that the faculty at the Athenaeum, like the scientific lecturers of the period, responded to the interests of their audience as much as they guided and refined those interests. Lectures at the Athenaeum were paid for

A typical scene from a late eighteenth-century class given at the Athenaeum in Amsterdam, which became the University of Amsterdam in 1877. (*Courtesy of the Museum of the University of Amsterdam***)**

by a subscription of about 30 guilders a year for a course and were even, on occasion, given purposefully at midday hours, when the Amsterdam stock market was closed. They were remarkably sophisticated in mathematics and astronomy, as notes taken by students at the time confirm (see illustration), but also by the end of the century concerned with industrial applications. It may be said that the Dutch Revolution (both in 1787 and 1795), from the perspective of widespread scientific education, reversed a trend of relative backwardness that had plagued the republic in the middle decades of the eighteenth century. To that extent, and in an attenuated sense, the revolution paved the way educationally for the Dutch Industrial Revolution of the mid-nineteenth century. Perhaps of greatest significance in regard to industrialization were the number of societies that became commonplace after the revolution which advocated utility and application.

From the early 1800s onward a new generation of school books also emphasized basic scientific education for both boys and girls.[64] In the Dutch Republic the sciences needed to promote commercial capitalism, in particular astronomy and meteorology, were very gradually joined by applied mechanics (and also chemistry) necessary for industrialization. This transformation began only very late in the eighteenth century, and those who effected it looked self-consciously at the British example, in both political and intellectual matters. To achieve an industrial vision, the scientific reformers believed, required a revolutionary displacement of the old elite such as they sought to effect in the 1790s and beyond. In 1800 the new revolutionary government made a vast survey of the condition of industry in the Republic, which revealed a rather appalling decline.[65] As political circumstances would have it, little was achieved prior to or immediately after 1815 in reversing that decline.[66] Scientific education aimed at industrial application could not in itself make an industrial revolution; but without it, systematic and sustained industrial development seemed unlikely.

BELGIUM

This pattern, whereby we see a link between sustained inquiry of a mechanical sort among entrepreneuring elites and early industrialization, would seem to hold for at least the important areas of industrialization in the southern Netherlands, for example, in the area around Charleroi. There, as previously mentioned, the Austrian government promoted manufacturing in the countryside, where an impoverished peasantry, unprotected by guilds, could easily be proletarianized.[67] But in the province of Liège, under the control of its bishop, that sort of imperial intervention was not possible. Among entrepreneuring manufacturers in that province we can observe a sustained agitation for mechanical and technical education, and this *esprit* was, as in England and The Netherlands, also associated with enlightened reform. Predictably, Freemasons were active in this scientific movement. In this enlightened propaganda the cause of industrialization was equated with the highest ideals of social utility.[68]

Indeed throughout the province of Flanders in the period after 1770, and at precisely the time when the first steps toward industrialization were being taken, we can observe the presence of a significant interest in mechanics and technology. The Flemish journal of the period, *Vlaemschen Indicateur,* reflects that interest on the part of literate elites also eager to promote the reforming policies of the Austrian monarchy.[69] Its ministers consistently used the new science and the establishment of scientific academies as a stick with which to beat the clerically controlled universities. The ideology of reform, and with it the promotion of industry through scientific inquiry, fitted well with the imperial need of the Austrians to overcome the localized interests of the indigenous and old Flemish aristocracy and clergy. The evidence from as early as the 1760s shows officials of the Austrian government intimately involved in the industrial process, in particular the nascent chemical industry, where they fostered research, lent money to entrepreneurs, and licensed their factories. The government's day-to-day involvement in the industrial process rewarded the indigenous entrepreneur and attracted foreign projectors, who brought with them new scientific knowledge, frequently from England.[70] In western Europe the enlightened advocacy of scientific education cannot be disassociated from the mentality of industrialization even where, as in France, the actualization of that social and economic process was for a variety of reasons delayed.

FRANCE

Despite the force of the Enlightenment in select French circles, industrialization on any significant scale did not occur there until the early nineteenth century. Of course within its scientific community, particularly but not exclusively when influenced by Newtonianism, the implications of applied mechanics were readily perceived. As Dutch reformers such as van Swinden noted later in the century, the French mechanist Jacques Vaucanson had attempted in the 1740s to establish factory production of a mechanical sort in the silk industry decades before Richard Arkwright founded his cotton spinning mill in Derbyshire.[71] There were French chemists of the early eighteenth century who knew that their science should be applied and who wanted the state to intervene and assist in that process. The vision of such men was quite simply industrial and included the training of workers whose skills would facilitate the entrepreneurs, who would in turn profit from the chemical applications.[72] And not least, the French scientific lecturer of the mid-eighteenth century, the abbé Nollet (1700–1770), was probably the most important itinerant promoter of the new science, complete with mechanical applications, on the Continent.

Predictably, Nollet learned his techniques of demonstration in the 1730s from s'Gravesande and the Dutch Newtonians. Thereafter he opened his *cours de physique* in Paris, a lecture series that he eventually took to the French provinces, the Low Countries, and Italy. This series was perhaps the most popular ever given on the Continent, and Nollet's fame came to rest partly on his electrical experiments, which astonished and delighted his audiences. Indeed popular enthusiasm for electrical effects cannot be ignored as one of the stimulants that enticed the eighteenth century general public to an interest in the new science, not least because it was believed that electricity possessed medicinal value and could cure everything from tumors to the gout. Yet for all of Nollet's importance—which rivaled that of Benjamin Franklin as an electrical experimenter—his course of physics was grounded firmly in the practical uses of the new science.

Like his British counterparts, Nollet had to know the interests and limitations of his audience. He eschewed complicated mathematical applications, provided a glossary of terms for his readers, and in general avoided metaphysical or physico-theological questions in favor of practical examples to illustrate the "mechanism of the universe." In this last aspect his lectures are representative of that general turn away from constant attention to religious questions, a shift clearly visible in scientific lectures given from the 1720s on both sides of the Channel. In concentrating on the useful, Nollet claims that he is catering to public taste; and to accommodate that interest he is using machines to illustrate the general principles of the new physics.[73] In the first instance, however, Nollet concentrates on basic chemistry; how to dissolve metals, such as gold coins, how to use glues in porcelain making, how to use nitric acid to dissolve iron filings, the techniques of dying cloth and paper—in short, the chemistry useful in trade and hand manufacturing.[74] The general laws of physics, such as inertia and resistance, are explicated verbally as well as illustrated by the impact of moving balls of lesser and greater size. Once these general principles are established, the mechanical lectures embark on explanations of how those laws may be employed "to the greatest advantage."[75] Here much is made of windmills for grinding, or pumps that raise water "for our use or for the decoration of our gardens," or vehicles for transportation, or levers and pulleys for architecture and navigation—all to be constructed not by simple "machinists" but by true mechanical philosophers. That such sophisticated machines can replace human labor and consequently save money is made quite obvious.[76] The approach taken by Nollet in his lectures could be described as proto-industrial, rather than directly industrial, in that little is made of the actual uses of mechanical devices in coal mining, water engineering, or manufacturing.

What is important to realize about the lectures of Nollet and the other French popularizers of the new science is that they provided the French elite with an alternative to the relative scientific backwardness of the University of Paris. Cartesianism was accepted there only in the 1690s, although it remained controversial in the eyes of the church (and the state) well into the 1720s. The first Newtonian lectures at the university were in the 1740s, and Nollet himself received recognition from its officialdom only in the 1750s.[77] If we contrast this pattern with natural philosophical teaching in the British or Dutch universities, or even in the provincial Dissenting academies in England by the

1740s, it is clear that a generation or more of French university students did not have access to knowledge directly useful to the process of industrialization. This is not to denigrate the large number of provincial literary and natural philosophical societies that spread all over France in the course of the century and in consequence made scientific knowledge available on an unprecedented scale. Yet aristocratic domination in those societies hardly permitted the kind of gentlemanly zeal for practical science that we see in late eighteenth-century Derbyshire or Birmingham (see pp. 164–166).

Far too much can be made of the contrast between French and British economic development in the course of the eighteenth century. Recent studies emphasize French progress, particularly in agricultural reform. Not least, there are no "rules" about when or how a country "should" industrialize. Yet that sort of enlightened relativism, found among present-day social scientists, does not reflect what later eighteenth-century contemporary reformers would have said. In 1793, at the height of the French Revolution, the Jacobin* Convention (or parliament) abolished the French scientific academies inherited from the old order, both in Paris and in the provinces. Two years later, it is true, the Paris Academy originally founded by Colbert in the 1660s (see pp. 63–65) was revitalized, reformed, and renamed; but its personnel was now quite different. We may well ask how and why that happened, why a revolutionary government, however wrong-headedly, sought to abolish academies we might associate with enlightened progress.

There can be no question but that from the time of Colbert onward, the French monarchical government showed a marked interest in science and its application. In the 1750s this interest focused on steam-powered boats, largely for military use; in the 1770s and 1780s encouragement was given to the invention of mechanical devices for agricultural application.[78] These efforts to introduce "scientific farming" were very extensive and reflected the highest ideals of enlightened absolutism as found in the decades prior to the French Revolution.[79]

The idealism behind those efforts was partly Baconian and partly a reflection of the secular idealism so commonplace among

---

*See glossary of terms.

the educated elites both aristocratic and nonaristocratic of the eighteenth century. Both they and the royal government supported scientific inquiry chartered or licensed by the crown. One of the major philosophes of the 1770s justified that linkage between absolutism and scientific inquiry in language that harks back to the debates of the early seventeenth century on the role of science within the state (see pp. 29–31). In urging the Spanish monarchy to institute an academy in its scientifically backward country, Condorcet, a leading philosopher of empiricism, explained that these academies are "an advantage for the monarchical state." His reasoning is as follows: "In a republic all citizens have the right to meddle in public affairs . . . but it is not the same in a monarchy. Those whom the prince appoints have the sole right to meddle." But for men who have a need to agitate and who cannot abide the inactivity forced on them by the nature of the monarchical state "the study of science can only represent . . . an immense vocation with enough glory to content their pride and enough usefulness to give satisfaction to their spirit."[80] For such men academies of science are needed, or so the argument went.

Of course, other arguments of a less overtly political nature were also routinely offered by the enthusiastic supporters of the new French academies. In 1781 the secretary of the Paris Academy expressed both his nationalism and his enlightened liberalism when he presumed that the other European academies "owe almost all their existence to the noble emulation and mass of enlightenment that the work of the Paris Academy of Science has spread throughout Europe."[81] Had he just said "France" there might have been considerable truth to it. The Paris Academy—which permitted only Parisians to join but excluded members of religious orders, such as the Jesuits—maintained a very high standard of original scientific inquiry throughout the century.[82] Many of the French provincial academies sought to imitate it. Their membership was overwhelmingly dominated by nobles, lawyers (many of whom worked with the nobility "of the robe," who were judges), and high clerics, who in the final years of the old order met together "in search of prestige and believing that progress [would result] from their collective reflection on new ideas."[83] They did everything from sponsor public lectures to become, later in the century, increasingly interested in technology, agriculture, and commerce. Yet in 1793 the revolutionary government took its vengeance on the academies, not on their ideals or on science per se, but on their personnel. The Paris Academy of Sciences lost nearly one half of its members as a result of the Reign of Terror;[84] the provincial *noblesse* were equally detested, if not persecuted.

Prior to the Revolution there had been a revival of philosophical naturalism of a materialistic sort. Some of its devotees once again dabbled in the mystical aspects of eighteenth-century science, for example, electrical cures performed by magicianlike healers. In that revival we see a profound disillusionment with establishment science, with the austere and rationalistic academicians and their private pursuit of scientific inquiry.

Yet at the Revolution, the science that triumphed more closely resembled engineering than it did magic.[85] *L'Ecole polytechnique* founded in 1794 embodied the ideals of a revolutionary vision of science, of its "power to change the world."[86] Its founders wanted nothing less than a school for the science of the revolution.[87] They ignored the universities, which they regarded as moribund; they closed the academies and sought instead to reeducate teachers and hence the young. In the final analysis they embraced an essentially industrial vision of the power of science to transform society and nature. A generation after his English counterpart of the 1760s and 1770s, the French civil engineer came into his own—not to displace his military counterpart (science in this period never abandoned the warmaking needs of the state) but to complement him in the new national state created by the Revolution.

In this abrupt turn toward industrialization one aspect of the Enlightenment ideals inherited from the old order now took preeminence over all others. Among the Parisian philosophes, especially those of bourgeois origins, there had been a marked interest in applied mechanics of the sort popularized by Desaguliers and Nollet. The greatest project of the Enlightenment—in terms of scope, size, and personnel—had been Diderot's *Encyclopédie,* which began to appear in 1751. Probably 25,000 copies of it circulated before 1789 and the outbreak of the Revolution. Its pages are filled with drawings and descriptions of mechanical inventions and devices. Its inspiration was Baconian; Diderot and his collaborators adored the new science and the promise it held to transform man's estate. As he put it, "Men struggle against nature, their common mother and their indefatigable enemy." In a utopian work intended to inspire the Russian monarch to establish the most modern of universities, Diderot urged that mechanics be the first science to be studied because it is "la science de première utilité."[88] The revolutionary instructors at *L'Ecole polytechnique* would have agreed.

This is not to suggest that prior to the French Revolution there had been a massive backwardness in mechanical knowledge among all segments of the French elite. By far the most scientifically literate of that earlier period were, however, the military engineers.[89] The preponderance of the state and the army in the area of technical and mechanical education naturally meant that their interests would be served before that of society's. That the new mechanical knowledge was most systematically exploited in the service of state-run projects, not least in the making of war but also in agricultural improvement,[90] apparently stifled the development of civil engineering relative to its development in Britain. The tendency to make science a creature of the state may have been further strengthened by the exclusivity of the engineering schools, which prior to the French Revolution consistently chose men of aristocratic birth for places in their classes.[91] In them, incidentally, the abbé Nollet's lectures were the standard text. Throughout the eighteenth century these French technician/scientists sought government patronage and the prestige that went with it.

There are moments in any survey of the social relations of eighteenth-century European science when two patterns seem most prominent: the French, where scientists in the first instance serve the state, and the British, where they service the needs of entrepreneurs. In neither case did the needs or interests of the majority of the people figure prominently.

The absence of a large standing army in mid-eighteenth-century Britain and of the concomitant necessity of channeling mechanical knowledge and talent in its service may have been significant for the development there of a cadre of civil engineers and scientific lecturers eager to find employment in whatever capacity. That they disseminated scientific knowledge on such a wide scale stands in contrast to the less commonplace character of that knowledge even in the most highly literate areas of western Europe, especially in The Netherlands but also in France.

ITALY

Where literacy, however, was weak and the power of censors strong the dissemination of the new science was infinitely more sporadic than in France or the Low Countries. In Italy, where once Galileo had captured the attention of both the elite and the censors, the new science of Gassendi, Descartes, and finally Newton held a tentative claim to allegiance among select circles in Rome, Naples, and Turin. In Rome, the city of the Inquisition, a circle of Gassendian atomists had met in the mid-seventeenth century, and for a brief time an academy there under the direction of Giovanni Ciampini dedicated itself to Galilean experimentalism and the study of Cartesian metaphysics.[92] In the late seventeenth century the intellectual crisis that afflicted much of western Europe (see Chapter 4) was also felt in Italy, and out of it came the linkage between science and heterodoxy. The search for philosophical liberty among scientifically minded Italian intellectuals in turn galvanized the Inquisition "against mathematics and physico-mathematics" because they were seen as pernicious "to the sincerity of religion."[93] Yet for all the dangers attached to the study of the new science, its penetration south of the Alps was real and durable.

The works of Robert Boyle made their way south as did various visiting British Newtonians. By 1707 Newton's *Optics* and *Principia* were the subject of avid discourse, and the polemic against Cartesianism had begun. The link between Newtonianism and Galilean mechanics was readily perceived, and not surprisingly it was an Italian engineer of Naples, Celestino Galiani, who made a major contribution to the formation of a Newtonian school in Italy.[94] Interest in the rationalization of navigation and agriculture, rather than industrial application, characterized the scientific *esprit* of these Newtonian circles. Not surprisingly they were also in close contact with the first generation of Dutch Newtonians.

An Italian edition of Francis Hauksbee's London lectures, *Physico-Mechanical Experiments on Various Subjects* (London, 1712)—

one of the first of those series of public lectures so central to this process of dissemination—appeared in Florence in 1716. This exposition set the tone for a scientific empiricism that set Italian Newtonians against Cartesians and scholastics for much of the century.

The Enlightenment in Italy made those polemics central to its concerns, and the Italian Newtonian F. Algarotti[95] sought to enlist educated women in the enlightened camp. His *Newtonianism for the Ladies* (1737), published in an Italian edition emanating from Milan, became the most widely read and translated general explication of the new science in the century. It may be seen as a bold appeal to enlist women against both the church and the Inquisition. Throughout the eighteenth century various scientific lecturers—the abbé Nollet and Benjamin Bosma, for example—reached for the support of that new segment of the literate population, not to offer them full membership in the scientific community but to enlist them as its passive supporters. It is yet to be determined how women responded in their own discourse to that appeal, whether they made use of natural philosophical arguments to criticize the inequity of their social place in every European society—and if so, how.

Newtonianism permitted liberal Italian Catholics to formulate a moderate and enlightened religiosity, indebted in part to the early Boyle lectures of Clarke and others (see pp. 95–96) that offered a *via media* between the materialism of the radical Enlightenment and the scholasticism advocated by the official church. In the face of material and social conditions totally unsuited to the promotion of industry, Italian Newtonians such as Antonio Genovesi concentrated their energies on the reorganization of the schools and academies. In Naples, one of the centers of the Italian Enlightenment, they sought nothing less than the modernization of their society and culture. Genovesi attacked the near feudal conditions that prevailed in the countryside and sought through the new science of economics to address the problems of poverty and agricultural backwardness.[96] The integration of science into Italian society may have produced a more immediately humane response to social problems than that found in either the British or French models. Yet it should be noted that in every European society a scientific approach to agriculture gained acceptance in the course of the eighteenth century and contributed significantly to the elimination of food shortages in major areas of western Europe.

Perhaps one of the most remarkable examples of the enlightened implantation of science occurred in Turin in the northern province of Piedmont. There in 1757 its aristocratic ruler simply began a new scientific academy where none had existed before. It quickly moved to the vanguard of contemporary European science, both pure and applied. Its proceedings, and those of the laboratory established to serve the needs of the army, display a remarkable interest in applied mechanics of an industrial sort. In this Piedmontese enlightenment, fostered by an absolute ruler, aspects of the modern relationship between science and the state appear with an uncanny prescience. Reform and improvement through science, progress, and liberalism are inextricably linked with war and warmaking. One out of fifty Piedmontese were involved in war or warmaking; the laboratories belonged to the technician/scientists of the army. The scene conjured up in our imagination of those decades looks toward the state-sponsored industrialization of the nineteenth century, toward the military-industrial complex of the twentieth.[97] We should not view the past with such "Whiggish" spectacles; yet at moments it is difficult to remember that we are wearing them.

THINKING MECHANICALLY

The ability to think mechanically—that is, scientifically, in the modern meaning of that word—permeated western European society selectively in the course of the eighteenth century.[98] In marginally literate segments of the western European population, and in certain areas of eastern Europe, that penetration occurred only in the nineteenth and twentieth centuries. In 1787 one of the first flights of a man-made balloon occurred some twelve miles outside Paris. When the balloon came to rest, frightened peasants mistook it for the moon falling; they attacked the object and badly damaged it.[99] In the late eighteenth century the Russian government attempted to import many of the mechanical devices developed in the West in the course of that century. British engineers were enlisted in the work of canal building and brought with them models of mechanical devices, not least of all the steam engine. When in 1780 these models were shown to older members of the military engineering corps—that body supposedly most educated in mechanical principles—some of them simply did not understand how such a machine could operate.[100]

Relatively sophisticated mechanical knowledge had to be a part of one's mental world before such mechanical devices could be invented and, more to the point, exploited. Where that knowledge was widespread, and where capital, natural resources, and exploitable labor were also present, the results of that coincidence transformed both nature and society, creating in its wake the modern industrial world.

NOTES

1. Ester Boserup, *Population and Technology* (Oxford: Blackwell, 1981), p. 4.
2. D. S. L. Cardwell, *The Organization of Science in England* (London: Heinemann, 1972), pp. 17–18.
3. See most recently, Thomas L. Hankins, *Science and the Enlightenment* (Cambridge: Cambridge University Press, 1985); and the magisterial study of Peter Gay, *The Enlightenment: An Interpretation. The Science of Freedom* (London: Wildwood House, 1973).
4. For a sophisticated statement of that lead, see G. Timmons, "Education and Technology in the Industrial Revolution," *History of Technology,* vol. 8 (1983), pp. 135–149. For a clear statement of how the "new" economic history discounts the entrepreneur, see Clive Trebilcock, *The Industrialization of the Continental Powers, 1780–1914* (London: Longman, 1981), p. 141; cf. pp. 63–65 on the critically important role of science and technology to late nineteenth-century German industrial development.
5. G. Vanpaemel, "Rohault's *Traité de Physique* and the Teaching of Cartesian Physics," *Janus,* vol. 71–74 (1984), pp. 31–40.
6. Quoted in C. R. Boxer, *The Dutch Seaborne Empire 1600–1800* (London: Hutchinson, 1965), p. 271. On navigational technology see C. A. Davids, *Zeewezen en Wetenschap. De wetenschap en de ontwikkeling van de navigatietechniek in Nederland tussen 1585 en 1815* (Amsterdam: De Bataafsche Leeuw, 1986). I wish to thank Dr. Davids for his helpful comments.
7. Ijsbrand van Hamelsveld, *De zedelijktoestand der Nederlandsche natie, op het einde der achttiende eeuw* (Amsterdam, 1791), p. 285; see also p. 244, where he calls for taking uncorrupted youths (from north Holland) and educating them "in art or science."
8. See Thomas A. Ashton, *Iron and Steel in the Industrial Revolution* (Man-

chester: Manchester University Press, 1963), pp. 42, 60, on the importance of inventions; note early eighteenth-century Dutch superiority in the art of iron casting.

9. As stressed, for example, by Phyllis Deane, "Industrial Revolution in Great Britain," in Carlo Cipolla, ed., *The Emergence of Industrial Societies* (Hassocks, Sussex: Harvester Press, 1976), p. 177, where, however, technological know-how in the Dutch republic (p. 174) is vastly exaggerated. For a good summary of the various Dutch contributions to this question, see J. G. van Dillen, "Omstandigheden en psychische factoren in de economische geschiedenis van Nederland," in *Mensen en achtergronden* (Groningen: Wolters, 1964), pp. 53–79.

10. A. Rupert Hall, "Further Newton Correspondence," *Notes and Records of the Royal Society of London,* vol. 37, no. 1 (1982), p. 32. I owe the point about Pitcairne to Anita Guerrini.

11. J. L. Heilbron, *Electricity in the Seventeenth and Eighteenth Centuries: A Study of Early Modern Physics* (Berkeley: University of California Press, 1979), p. 142.

12. Ibid., p. 26. He speaks of the *Principia.*

13. J. T. Desaguliers, *De Natuurkunde uit Ondervindingen* (Amsterdam: Isaak Tirion, 1751; first edition 1736). Cf. Edward G. Ruestow, *Physics at Seventeenth and Eighteenth Century Leiden: Philosophy and the New Science in the University* (The Hague: Nijhoff, 1973), pp. 143–144; cf. C. de Pater, *Petrus van Musschenbroek (1692–1761) een Newtonians natuuronderzoeken* (Utrecht: Elinkwijk, 1979).

14. See D. van der Pole, "De introductie van de Stoommachine in Nederland," in J. de Vries, ed., *Ondernemende Geschiedenis* (The Hague: 1977).

15. Royal Library, The Hague, MS 128 B. 3., s'Gravesande MSS. Cf. J. N. S. Allamand, *Catalogus van eene aanzienlijke Verzameling van allerleije . . . Instrumenten* (Amsterdam, 1788), which includes a list of s'Gravesande's instruments, among them copies of windmills and water mills, electrical devices, etc.

16. Royal Society, MS 702, e.g., s'Gravesande, Justus van Effen, Sallengre, St. Hyacinthe, William Bentinck. On Sallengre and Newton, see A. Rupert Hall, "Further Newton Correspondence," p. 26.

17. University Library, Leiden, Marchand MS 2, 15, 7bre, 1723, from Surinam; Jac. de Roubain to P. Marchand: "Vous pourrez en être plus particulièrement informée le plan que j'ai ici joint, et si vous vouliez abjurer le Newtonniste je suis aussi puis d'abjurer le Carthesianisme."

18. A. C. de Hoog, "Some Currents of Thought in Dutch Natural Philosophy," (Ph.D. diss., Oxford, 1974), p. 295. On Fahrenheit, see University Library, Leiden, MS. BPL 772; and Pieter van der Star, ed. and trans., *Fahrenheit's Letters to Leibniz and Boerhaave* (Amsterdam: Rodopi, 1983), p. 13.

19. Heilbron, *Electricity,* p. 159.

20. For an excellent overview of the impact of scientific education in Britain, see I. Inkster, "The Public Lecture as an Instrument of Science Education for Adults—The Case of Great Britain, c. 1750–1850," *Paedogogica historica*, vol. 20, (1981), pp. 80–112.

21. For a list of these societies, see J. H. Buursma, *Nederlandse Geleerde Genootschappen opgericht in de 18ᵉeeuw* (The Hague: Discom, 1978); cf. James E. McClellan III, *Science Reorganized: Scientific Societies in the Eighteenth Century* (New York: Columbia University Press, 1985), pp. 9–10.

22. For the transactions of this society, see *Verhandelingen uitgegeeven door de Hollandse Maatschappij der Wetenschappen, te Haarlem*, vol. 1 (1754) to vol. 11. Cf. MSS of the society, at its offices in Haarlem, "Notulen 1752–67"; see also R. J. Forbes, ed., *Martinius van Marum, Life and Work* (Haarlem, 1969); and J. A. Bierens de Haan, *De Hollandsche Maatschappij den Wetenschappen, 1752–1952* (Groningen: Willink, 1977).

23. W. W. Mijnhardt, "Het Nederlandse genootschap in de achttiende eeuw en vroege negentiende eeuw," *De Negentiende Eeuw*, vol. 2 (1983), pp. 79–81.

24. J. van den Berg, "Eighteenth Century Dutch Translations of the Works of Some British Latitudinarian and Enlightened Theologians," *Nederlands archief voor Kerkgeschiedenis*, n.s., vol. 59, no. 2 (1979), p. 198, on Samuel Clarke's influence. Cf. T. Dekker, "De popularisering der natuurwetenschap in Nederland in de achttiende eeuw," *Geloof en Wetenschap* (1955), pp. 173–188; R. Hooykaas, "De natuurwetenschap in 'de eeuw den genootschappen,' " in H. A. M. Snelders and K. van Berkel, eds., *Natuurwetenschappen van Renaissance tot Darwin* (The Hague, 1981), pp. 131–167.

25. Rijksarchief, Arnhem, MSS of the Academy of Harderwijk, no. 154, 153, 155, 156, 157, 141.

26. Rijksarchief, Arnhem, MSS of J. van Leeuwen, no. 5 and 6; note praise of Freemasons (no. 6, fol. 10 ff.).

27. Willem Frijhoff, "Deventer en zijn gemiste universiteit, Het Athenaeum, in de sociaal-culturele geschiedenis van Overijssel," *Vereeniging tot Beoefening van Overijsselsch regt en geschiednis, Verslagen en Medeelingen*, vol. 97 (1982), p. 71.

28. Thomas Schwenke, *Noodig bericht over de Inventinge der Kinderpokjes* (The Hague, 1756), p. 15; he was able to inoculate only 41 prominent citizens in a city of approx. 35,000.

29. G. J. Hollister-Short, "The Introduction of the Newcomen Engine into Europe," *Transactions of the Newcomen Society*, vol. 48, (1976–1977), pp. 11–22. On the efforts of the Austrian officials to promote the new science in the Low Countries, see Ghislaine de Boom, *Les Ministres Plenipotentiaires dans les pays-bas autrichiens principalement Cobenzl* in *Academie Royale de Belgique, Memoires*, 10th ser., vol. 31 (1932), pp. 59–61 and 235–236, on science used as the weapon to reform the universities.

30. K. W. Swart, "Holland's Bourgeoisie and the Retarded Industrialization of the Netherlands," in F. Krantz and P. Hohenberg, eds., *Failed Transitions to Modern Industrial Society: Renaissance Italy and Seventeenth Century Holland* (Montreal: Interuniversity Centre for European Studies, 1975), pp. 44–45; see entire volume.

31. I. K. van der Pols, "Early Steam Pumping Engines in the Netherlands," *Transactions of the Newcomen Society,* vol. 46–47 (1973–1976), pp. 13–16. See also Peter Mathias, "Skills and the Diffusion of Innovations from Britain in the Eighteenth Century," *Transactions of the Royal Historical Society,* vol. 25 (1975), p. 99, where we also learn that Dutch artisans were prominent in technology transfer, but to Spain and Russia (p. 94). On use of the steam engine by the Austrian government, see M. Teich, "Diffusion of Steam-, Water-, and Air-Power to and from Slovakia During the 18th Century and the Problem of the Industrial Revolution," *Colloques Internationaux, Centre National de la Recherche Scientifique,* no. 538.

32. For a survey of the market for science books, see A. J. Meadows, ed., *Development of Science Publishing in Europe* (Amsterdam: Elsevier, 1980). For a general look at Dutch publishing in science, see D. Bierens de Haan, ed., *Bibliographie Neerlandaise historique-scientifique. Des ouvrages importants dont les auteurs . . . 16ᵉ, 17ᵉ, et 18ᵉ siècles* (Rome, 1883; reprinted 1960).

33. For example, see W. A. Speck, "Politicians, Peers, and Publication by Subscription 1700–50," in Isabel Rivers, ed., *Books and Their Readers in Eighteenth Century England* (Leicester: Leicester University Press, 1982), p. 64, for merchants and M.P.s heavily subscribing to H. Pemberton's *View of . . . Newton's Philosophy* (London, 1728); but also see entire volume.

34. On this whole question of decline, see J. de Vries, *De economische Achteruitgang der republiek in de achttiende eeuw* (Leiden: Kroese, 1968); and J. Mokyr, *Industrialization in the Low Countries, 1795–1850* (New Haven: Yale University Press, 1976). On the general concern within the republic at the time, see G. van der Meer, "Prijspenningen van Nederlandsche geleerde genootschappen in de achttiende eeuw," *Documentatieblad werkgroep achttiende eeuw,* vol. 15 (1983), pp. 1–20.

35. [Anon.], *Aanspraak gedann aan de Goede Burgeren, die tot Welzyn van stad en land, op den 9 Augustus 1748, op den Cloveniers Doelen vergadert zyn geweest* (Amsterdam, 1748), p. 1: "de Konsten en Wetenschappen zyn onbeloond van ons gevlooden; de Koophandel is haare Stief-Vaders ontvlugt; de Fabriquen, die onuitputbaare Goudmynen der Volkeren, en waarop deeze STAAT met regt zig voormaals dorft beroemen, en waarop dezelve is gevest, zyn naar andere Natien overgegaan."

36. Marten G. Bruist, *At Spes non Fracta. Hope & Co. 1770–1815: Merchant Bankers and Diplomats at Work* (The Hague: Nijhoff, 1974), p. 9.

37. University Library, Amsterdam, MS. X.B.1, "Leçons de Physique de Mʳ le Prof. Koenig qu'il a donne à la Haye, 1751–52," 348 ff. These

lectures were almost certainly for the circle around the Bentinicks and the court, given the opening remarks and the use of French.

38. Ibid., fols. 6–16.
39. Ibid., fol. 33; fol. 79 ff.
40. Ibid., fols. 234–50.
41. Ibid., fols. 139–59.
42. Royal Library, The Hague, MS 75. J. 63, "Leçons d'Arithmétique et d'Algebre a l'usage . . . Le Prince d'Orange," May 1759, fol. 34 ff.
43. J. F. Martinet, *Katechismus der Natuur,* 4 vols. (Amsterdam, 1777); and for a popular English translation intended for children, see *The Catechism of Nature for the Use of Children by Doctor Martinet,* trans. John Hall (London, 1800), especially pp. 69–71.
44. See Giles Barber, "Aspects of the Booktrade Between England and the Low Countries in the 18th Century," *Documentatieblad werkgroep achttiende eeuw,* no. 34–35 (1977), pp. 47–63; and Robert Schofield, *Mechanism and Materialism: British Natural Philosophy in an Age of Reason* (Princeton: Princeton University Press, 1970), pp. 137–140, on B. Nieuwentyt's *The Religious Philosopher* (1718–1719) and its many English editions; the translator was Desaguliers, who compared the author to John Ray and William Derham. Cf. de Hoog, "Dutch Natural Philosophy," p. 295 ff. on Nieuwentyt.
45. Simon Schama, *Patriots and Liberators: Revolution in the Netherlands, 1780–1813* (New York, 1977), p. 50.
46. Dr. William's Library, London, Wodrow-Kenrick correspondence, MS. 24. 157, fol. 41; dated 1760.
47. Ibid.
48. See MS of Concordia et Libertate, Gemeente Archief, Amsterdam, P.A.9.1–10.
49. See A. J. J. Ph. Haas, "De Saturdagse Krans 1718–93. Een gezellige vereeniging van Amsterdamsche Regenten in de 18de eeuw," *Koninklijk Oudheidkundig genootschap Amsterdam,* vol. 77 (1934–1935), pp. 66–79.
50. B.L. MSS ADD. 6858, fol. 35, Elizabeth Montagu to A. Mitchell (in Berlin), March 6, 1767.
51. *De Koopman,* vol. 1 (1768), pp. 40, 332–333.
52. Ibid., pp. 335–336.
53. Ibid., vol. 4 (1773), p. 172.
54. See MSS of Felix Meritis, Gemeente Archief, Amsterdam, P.A.59. 19.
55. [Anon.], *Redenvoering over het algemeen nut der Wetenschappen, fraaije letteren en konsten . . . Felix Meritis* (1788); bound with J. H. van Swinden, *Redenvoering en aanspraak ter . . . inwijling van het gebouw der maatschappij Felix Meritis* (Amsterdam, 1789), pp. 29–30.
56. H. A. M. Snelders, "Het Department van natuurkunde van de Maatschappij van verdiensten Felix Meritis in het eerste kwart van zijn bestann," *Documentatieblad werkgroep achttiende eeuw,* vol. 15 (1983), p. 200.

57. Benjamin Bosma, *Gronden der Natuurkunde* (Amsterdam, 1764). The edition of 1793 states the author's pride at having continued this tradition of lecturing for so many decades. Concordia et Libertate gave money to the radical reformers in 1748.

58. Benjamin Bosma, *Redenvoering over de Wijsbegeerte* (Amsterdam, 1767).

59. Benjamin Bosma, *Redenvoering over de Natuurkunde* (Amsterdam, 1762), pp. 5–6.

60. Ibid., p. 8.

61. *Beknopte aanspraak, van den Heere Martinus Martens, uitgesprooken volpens jaarlykse gewoonte op den 6 Februari 1741* (Amsterdam, 1741), pp. 6, 12, 15, 17.

62. *Korte Beschrijving van de samenstelling en werking der Vuur of Stoommachine volg. Watt en Boulton. Met het rapport van J. H. van Swinden en C. H. Damen daarover* (1789); University of Amsterdam, Library, sign 473.A 13. Cf. H. A. M. Snelders, "Lambertus Bicker (1732–1801), An Early Adherent of Lavoisier in The Netherlands," *Janus,* vol. 67 (1980), pp. 104–122n. For another example of the link between industrial interests and the *patriotten* movement, see C. Elderink, *Een Twentsch Fabriqueur van de achttiende eeuw* (Hengelo: Broekhuis, 1977), pp. 73–74.

63. On the Athenaeum, see *Gedenkboek van het Athenaeum en de Universiteit van Amsterdam, 1632–1932,* Amsterdam, 1932. I am very grateful to Mrs. Feiwel for her assistance with these archives.

64. For example, *Van Vaderlandsche Mannen en Vrouwen uit de zuidelijke provincien: Een Schoolboek. Uitgegeven door de Maatschappij tot Nut van 't Algemeen* (Leiden, Deventer, and Groningen, 1828; many subsequent editions). On educational reforms after the revolution and the importance attached to science, see Aart de Groot, *Leven en Arbeid van J. H. van der Palm* (Utrecht: University of Utrecht, 1960).

65. "Journal der reize van den agent van Nationale economie der Bataafsche Republick," *Tijdschrift voor Staathuishoudkunde en statistiek,* vols. 18 and 19 (1859–1860).

66. E. H. Kossman, *The Low Countries, 1780–1940* (Oxford: Clarendon Press, 1978), pp. 97–100.

67. See Max Barkhausen, "Government Control and Free Enterprise in Western Germany and the Low Countries in the Eighteenth Century," in Peter Earle, ed., *Essays in European Economic History, 1500–1800* (Oxford: Clarendon Press, 1974), pp. 248–250. And Herve Hasquin, *Le 'Pays de Charleroi' aux XVIIe et XVIIIe siècles. Aux origines de la révolution industrielle en Belgique* (Brussels: Université libre de Bruxelles, 1971), p. 80; for interest in the Newcomen engine with a mastery of its operation, see pp. 138–139n.

68. D. Droixhe, "Noblesse éclairée, bourgeoisie tendre dans la principauté de Liège au XVIIIe siècle," *Études sur le XVIIIe siècle,* vol. 9 (1982), pp. 9–47, especially, pp. 24–31.

69. Herve Hasquin, ed., *La vie culturelle dans nos provinces au XVIIIe siècle* (Brussels: Credit Communal, 1983), p. 132–133.
70. Annette Andre-Felix, *Les débuts de l'industrie chimique dans les Pay-Bas autrichiens* (Brussels: Université libre de Bruxelles, 1971).
71. Siegfried Giedion, *Mechanization Takes Command* (New York: Norton, 1969), p. 35. [For a balanced account of Vaucanson, see Charles C. Gillespie, *Science and Polity in France at the End of the Old Regime* (Princeton: Princeton University Press, 1980), pp. 414–417.]
72. D. Todericiu, "Jean Hellot (1685–1766), savant chimiste, fondateur de la technologie chimique en France an XVIIIe siècle," *Comptes rendus du 105° Congres National des Societés Savants,* Caen, 1980, pp. 201–211.
73. Abbé Nollet, *Leçons de Physique experimentale* (Amsterdam and Leipzig, 1754), vol. 1, preface, pp. xxii–xxv.
74. Ibid., vol. 1, p. 44.
75. Ibid., vol. 3, pp. 1–5.
76. Ibid., p. 8.
77. L. W. B. Brookliss, "Aristotle, Descartes and the New Science: Natural Philosophy at the University of Paris, 1600–1740," *Annals of Science,* vol. 38 (1981), pp. 57–58, 67–68; cf. for a good general discussion, Henry Guerlac, *Newton on the Continent* (Ithaca: Cornell University Press, 1981).
78. Shelby J. McCloy, *French Inventions of the Eighteenth Century* (Lexington: University of Kentucky Press, 1952), pp. 30–31, 112–113.
79. R. Rappaport, "Government Patronage of Science in Eighteenth Century France," *History of Science,* vol. 8 (1969), pp. 119–136.
80. James E. McClellan, "Un Manuscrit inedit de Condorcet: Sur l'utilité des académies," *Revue d'histoire des sciences,* vol. 30 (1977), pp. 247–248; cf. Keith Baker, *Condorcet* (Chicago: University of Chicago Press, 1975), pp. 2–28, 401. For science in eighteenth-century Spain, see David Goodman, "Science and the Clergy in the Spanish Enlightenment," *History of Science,* vol. 21 (1983), pp. 111–140.
81. McClellan, *Science Reorganized,* pp. 9–10.
82. Heilbron, *Electricity,* pp. 115–117.
83. Daniel Roche, *Le Siècle des lumières en Province* (Paris: Mouton, 1978), vol. 1, p. 329.
84. Dorinda Outram, "The Ordeal of Vocation: The Paris Academy of Sciences and the Terror, 1793–95," *History of Science,* vol. 21 (1983), pp. 254–255.
85. John Hubbel Weiss, *The Making of Technological Man: The Social Origins of French Engineering Education* (Cambridge, Mass.: MIT Press, 1982), pp. 13–24.
86. Jean Dhombres, "L'enseignement des mathématiques par la 'methode révolutionnaire.' Les leçons de Laplace à l'Ecole normale de l'an III," *Revue d'histoire des sciences,* vol. 33 (1980), pp. 315–348.
87. Janis Langins, "Sur la première organisation de l'Ecole polytech-

nique. Texte de arrêté du 6 frimaire an III," *Revue d'histoire des sciences,* vol. 33, (1980), pp. 289–313.

88. Denis Diderot, *Oeuvres completes* (Paris, 1875), vol. 3: "Plan d'une université pour le gouvernement de Russie," p. 429, for "leur mère commune et leur infatigable ennemie"; and p. 457.
89. Charles C. Gillespie, *Science and Polity in France at the End of the Old Regime* (Princeton: Princeton University Press, 1980), p. 90.
90. R. Rappaport, "Government Patronage of Science in 18th Century France," *History of Science,* vol. 8 (1969), pp. 119–136.
91. C. Stewart Gillmore, *Coulomb and the Evolution of Physics and Engineering in Eighteenth Century France* (Princeton: Princeton University Press, 1971), pp. 12–14. In The Netherlands, too, military engineering was much more highly developed than was civil; see Harry Lintsen, *Ingenieurs in Nederland in der negentiende eeuw* (The Hague: Nijhoff, 1980), pp. 23–28. For a good illustration of the French "style" of scientific inquiry versus the British, see Richard Gillespie, "Ballooning In France and Britain, 1783–1786," *Isis,* vol. 75 (1984), pp. 249–268.
92. Vincenzo Ferrone, *Scienza, Natura, Religione. Mondo Newtoniano e cultura italiana del primo settecento* (Naples: Jovene, 1982), p. 4.
93. Ibid., p. 11.
94. Paola Zambelli, "Antonio Genovesi and Eighteenth Century Empiricism in Italy," *Journal of the History of Philosophy,* vol. 16 (1978), pp. 198–199.
95. Ferrone, *Scienza,* pp. 254–256.
96. Zambelli, "Antonio Genovesi," p. 208.
97. See Vincenzo Ferrone, "Tecnocrati militari e scienziati nel piemonte dell'antico regime. Alle origini della reale accademia della scienze di torino," *Rivista storica italiana,* vol. 96, no. 2 (1984), pp. 414–509. Note the presence here of Freemasonry.
98. For an excellent account of the pervasiveness of scientific culture in England, see Roy Porter, "Science, Provincial Culture and Public Opinion in Enlightenment England," *British Journal for Eighteenth Century Studies,* vol. 3, no. 1 (1980), pp. 20–46.
99. Shelby T. McCloy, *French Inventions of the Eighteenth Century* (Lexington: University of Kentucky Press, 1952), p. 13.
100. British Library, MSS ADD. 33, 564, diary of Samuel Bentham while in Russia, fol. 21. The machine shown was for driving piles. For a much more comprehensive treatment of Russian science than is possible here, see Valentin Boss, *Newton and Russia: The Early Influence, 1698–1796* (Cambridge: Harvard University Press), 1972.

CHAPTER 7

Science in the Industrial Revolution

In the first place, when the law of acquisition is established, it is impossible to restrict the amount. That must be left to a variety of personal qualities. . . . When you advert to the system I have laid down, that each person is but an engine in the great mechanism of circulation, you will be convinced, that the general plan, upon which wealth is distributed, is also the most salutary.

From *A Dialogue between a gentleman and a mechanic*, c. 1798

The mechanization of nature through the discoveries of the new science provided early industrialists with an arsenal of new knowledge, as well as with new metaphors of self-justification, which could be applied in the service of their economic interests. The image of the individual as an "engine in the great mechanism of circulation" rendered the disparities between rich and poor as simply the social consequences of the mechanical forces and laws at work in the natural order as they applied to the social order. The entire process of industrialization could therefore be seen— then and now—as the coming together of impersonal factors, necessities arising out of the application of surplus capital to raw materials, or the appearance, willy-nilly, of new technical inventions in an attempt to overcome low profit margins. Once begun these innovations became "cumulative" in number and effect, literally "self-sustaining,"[1] or alternatively, the effect simply of a neat interaction among a variety of factors: rising demand at home for consumer goods, higher wages, and the impossible burden faced by the entrepreneur of trying to render technical innovations unnecessary when the supply of skilled labor was always so precarious. But to these mechanisms of economic development—not in themselves unreal when used cautiously to offer general descriptions of long-term changes resulting in an overall transformation of society—we must now add another perspective.

THE APPLICATION OF MECHANICAL KNOWLEDGE

In this chapter we shall approach moments in the early industrialization of England, not as examples of these historical mechanisms so beloved by economic historians, but rather as instances of discrete choices and calculations made by merchants or entrepreneurs or public bodies and private associations, on the basis of the new scientific knowledge now readily available to them. Knowing the extent and nature of scientific education available to the literate classes by the middle of the eighteenth century (Chapter 5), can we see the effects of that education at work in society and industry? Can we show more precisely how access to mechanical knowledge could become a means, and a motive in itself, for seeking its application? A variety of projects, long regarded as central to the historical process of industrialization, in particular canal building and the use of steam power, will be examined in order to illustrate how scientific knowledge could be brought to bear in a variety of ways that directly affected profit and productivity. The contention here is that the long process of cultural integration through which science passed in the seventeenth and eighteenth centuries had created by the second half of the eighteenth century a new type of European. His signal characteristic was access to, and understanding of, the mechanical aspects of the new scientific learning. He is found most readily in England at the vanguard of industrial and commercial activity; but there and elsewhere he could also be a landed gentleman using his land in capital intensive ways and profoundly interested in agricultural improvement. He is, of course, to be found in many western European countries by this period, although not always with access in these places to the political power necessary to effect the changes he desired. That he can be located so conspicuously and so early in Britain—in the west country, in Derbyshire, and of course, in Scotland—must be acknowledged as one important piece in the puzzle of why England industrialized first.

Such a man (and almost exclusively the general mentality I am

describing was found among men and not women, with consequences I shall examine briefly in the epilogue) understood nature mechanically and wished to use that knowledge for his
financial profit and sometimes also for the general improvement
of society. He approached natural obstacles inhibiting transportation or manufacturing or mining with this mechanical knowledge of nature at the forefront of his thinking. Obviously this
does not mean that he thought less about labor costs, or com-
peted less ruthlessly with other entrepreneurs, or treated his
workers more or less harshly; but it does mean that he engaged
in economic activity armed with a new and compelling kind of
knowledge. He understood the ways things worked in the world,
and that knowledge gave him a consciously felt power which
sometimes also resulted in personal economic advancement in
excess of that achieved by his less knowledgeable competitors. At
the least this knowledge translated into a self-confidence that
applied the mechanical model of nature to society and assumed,
therefore, that profit based on the wage labor of others and on
the manipulation of markets was simply in the nature of things,
that order existed beneath the apparent disorder of self-interest
and market forces.

Curiously enough, the economic and social, as opposed to the
intellectual and cultural, explanations offered to explain the origins of the Industrial Revolution were first derived from the very
mechanical laws of the new science used by late eighteenth-
century entrepreneurs to justify their dominant place in society.
Those mechanistic explanations in turn adopted by the social
sciences have enjoyed a remarkable hegemony and tenacity in the
historical literature about industrialization. For a variety of motives many contemporary historians, whether liberal, conservative or Marxist, have been unable to shed the assumption, first
articulated by the ideologues who defended the capitalization of
agriculture and industry, that what the industrial capitalist was
doing was somehow inevitable, the result as it were of the laws
of nature experienced, but never thoughtfully understood or
purposefully applied, by their beneficiaries. Current-day social
historians, in particular, still rely almost entirely on statistical
models and on impersonal laws—such as the relationship between alterations in population growth and the appearance of
new productive modes—to explain the origins of industrializa-

tion.[2] Some of these laws have been derived from the writings of Karl Marx, perhaps the most astute and obsessive observer of industrialization to be found in his age. As we know, Marx, like a few of his present-day followers, shared in some of the more grandiose assumptions common to the early nineteenth century about the possibility of finding social laws as universal as the natural laws of science—the latter, and by implication the former, he presumed to be timeless and value free.

But for all his naiveté about the historicity of science, Marx himself can never be accused of ignoring culture, of a disinterest in the thought processes available to European elites at the early stages of industrialization.[3] To deny the importance of scientific knowledge as one of the pillars on which rested the power of that class he chose to call the bourgeoisie would have been for Marx, as a moral thinker, a dangerous myopia about their resourcefulness, and hence a profound disservice to the very proletariat he sought to liberate. To imagine that the scientific learning described in the previous chapters had little direct relevance to that historical development we call the Industrial Revolution is to ignore the rich historical evidence now available for the application of science to commerce and industry. The power inherent in scientific thought was seized upon, and applied, by late eighteenth-century entrepreneurs who had come to understand nature mechanically.

Historical explanations that exhibit simplicity and monocausality as their distinguishing characteristics are not only flawed, they are also frequently boring and ultimately irrelevant. Consequently, to emphasize in this chapter the importance of scientific thinking in the decision making of entrepreneurs, and then to assert the primacy of these mental operations over economic considerations or material circumstances, would be to err in the same monocausal way, only using a different set of simplistic assumptions about historical change, as do the many economic and social historians who emphasize material and impersonal forces as the critical, and indeed as the only, factors in early industrialization. A balanced account, one true to the complexity of human nature and hence history, would seek to show that certain moments of decision making—where the decision to proceed or not directly assisted or inhibited industrial development —were shaped by the availability of scientific knowledge, among other beliefs and judgments. Here we can do no more than present a few concrete examples to illustrate the generalization that to an extraordinary extent scientific knowledge had penetrated the thinking of literate Englishmen by the late eighteenth and early nineteenth centuries, and that such knowledge contributed directly to the process of industrialization, to creating the world in which we now live.

Frequently these scientifically educated gentlemen considered themselves to be as mechanically knowledgeable as the professional mechanists. For practicing engineers such as John Smeaton (see pp. 160–163) and his successors—William Jessop (whom he trained), Benjamin Outram, and John Rennie, whose canals and bridges revolutionized transportation in Britain—the scientific learning of the entrepreneurs offered immense opportunity for employment and even for business collaboration; but it was also the source of much irritation and conflict. Writing in the 1760s Smeaton complained of his entrepreneurial employers —that is, the members of the canal companies who paid his fees to draw up plans and then to explain and defend them in Parliament—because they so frequently interfered with the execution of these plans. Writing in the first instance of the foremen who supervised the on-site work of the navvies or canal diggers and who then reported back to the directors of the canal companies, Smeaton complained: "Not only all of the inferior departments are ambitious to be practical engineers," but "even members of the company *have a propensity that way* too; by which means *all becoming masters* . . . the parties interfering *suppose themselves competent* to become Chief Engineers." He ruefully observed that "they cannot have great length of experience [like himself] in conducting public works" and do not possess the degree of theoretical knowledge they imagine themselves as commanding.[4]

As must have been evident to Smeaton and his contemporaries, men and women are not born with the ability to conceive of nature mathematically and mechanically, nor with the ability to invent mechanical objects of anything but the most rudimentary simplicity. That seems so obvious, yet when considered in relation to the late eighteenth century it acquires a startling historical significance. In the world in which we live such deficits at birth are quickly remedied through daily experience of mechanical devices or their effects, and then, of course, through universal formal education in basic science. As a result it is extremely difficult, indeed it requires a leap of historical imagination, to conceive of the time when the mechanical understanding of nature was new and anything but commonplace, when its assumptions violated centuries-old explanations about nature that rested on nonmechanical beliefs. Imagine at that time being in the unique situation of Smeaton and his employers, of being a member of a small elite conversant with science—someone who had attended scientific lectures, who may even have been taught Newtonian science by a particularly advanced schoolmaster, or who had possibly become so passionately interested in scientific learning as to qualify, as did Smeaton, as a Fellow of the Royal Society. You would have been, and may have conceived of yourself as being, in possession of a new and powerful wisdom. Such scientifically knowledgeable men could be found in every town of any significance in late eighteenth-century Britain. They frequented literary and philosophical societies, attended scientific lectures, read scientific books, promoted new transportation schemes, joined agricultural societies, or even installed steam engines in their factories, often at considerable capital risk. When they thought about the natural world they saw it as measurable, as a set of push-pull interactions which could release power that could be maximized by its application to machinery. They thought of water, wind, hills, and valleys as places where canals might be built or steam engines used, provided the terrain could first be measured and the correct principles of leverage and pressure applied to regulate the flow of water or the power of an engine. And they thought of these man-made objects as beautiful in themselves, as aesthetically pleasing as well as useful and profitable.

When these entrepreneurs of transportation and industry put up the capital for these projects, their foremen in turn went out and hired unskilled or semiskilled workers to dig the trenches and tunnels for the canals or to feed coal into the burners of steam engines. This human element in the early Industrial Revolution was the least subject to the mechanical laws explicated by the scientific lecturers. As one engineer complained: "Stone, wood, and iron, are wrought and put together by mechanical methods; but the greatest work is to keep right the animal part of the machinery,"[5] that is, the workers. We know from other sources that these unmechanized men, who frequently lost their lives in the digging of tunnels or the mining of coal, would consult the position of the stars before they began their arduous and sometimes dangerous work.[6] For them astrological calculations and magical beliefs were appropriate to the uncertainty of their livelihood, even of their lives; they were more meaningful perhaps than any other form of natural explanation. And certainly if a worker was illiterate or could not afford scientific books, and the majority of workers in this period could not, he also had no access to an alternative rendering of the relationship between the natural order and everyday events, and the new mechanical learning.

When the promoters of transportation and industrial improvements consulted their communities to gain their support or financial backing for a new canal, predictably they did not consult these workers. Rather they appealed to men with some capital who were also literate and hopefully knowledgeable in mechanical matters so that they could understand the merits, if not the actual engineering details, of the proposed plans. It is possible to distinguish levels of scientific learning among such middling sorts of men and to compare their understanding of the world with that possessed by engineers or fellows of the Royal Society. Such a microscopic examination of individuals poised at the beginning of a world where the signs of industrialization—the factories, canals, harbors, bridges, and steam engines—were now apparent should reveal the multiplicity of natural systems of explanation still prevalent at that moment and illustrate the singular relevance of mechanical explanations for the promoters of commerce and industry.

BRISTOL: AN EARLY EXAMPLE OF THE APPLICATION OF MECHANICAL SCIENCE

The city of Bristol in the west of England, the commercial metropolis for the west country and center of the Atlantic trade, provides such a microcosm for our examination of the application of applied mechanical science.[7] In this city (population 60,000 by the 1760s) many of the preconditions of industrialization were already present. The west country was rich in mineral deposits, mining was commonplace, advanced forms of iron production were already widespread,[8] while surplus capital needed for investment came to Bristol's merchants from the Atlantic trade—especially in slaves, tobacco, and sugar. Yet despite these signs of early or proto-industrialization, Bristol and the west country in general would ultimately, by 1810, lose ground to its northern rival, the growing port city of Liverpool and its environs. Indeed the response of Bristol's merchant community to that threat provides an occasion for discovering the degree of mechanical knowledge to be found among its commercial elite— the leading merchants who often invested in industrial ventures and who exercised an inordinate amount of authority over its day-to-day political life and town government.

The New Science in Bristol

Bristol is all the more interesting for illustrating the widespread availability and use of scientific knowledge because it possessed no scientific society of its own. Indeed one of the leading scientific minds among the city's urban gentry, Richard Bright, had to use the services of the distant Manchester Literary and Philosophical Society as he struggled to convince his mercantile colleagues of the necessity for a new floating harbor to be built with the best engineering advice of the day. The evidence for the varieties of scientific learning found in Bristol and its environs comes therefore not from a single source, such as exists for Manchester, Derby, or Spalding (see pp. 152–160), but from a variety of sources, and as a result it is all the more fascinating.

For instance, when the prominent scientific lecturer of the 1760s and 1770s, James Ferguson, F.R.S., gave his course on mechanics, hydrostatics, and hydraulics in Bristol, he dined with a member of his audience—a locally prominent accountant, part-time doctor, and student of natural philosophy, William Dyer.[9] Dyer was also an electrical practitioner, well read in the latest experimentation, who used electrical shocks, as was common at the time, in his medical practice. He applied the technique to a wide range of diseases and ailments—from rheumatism, the gout, and "lombago," where it was apparently helpful, to consumption and deafness, where it seems to have been quite useless. This man of the new science, on the very evening he dined with Ferguson, also visited his close spiritual friend, one Rachael Tucker, a prophetess who "possessed the intimate worship of God."[10] Dyer was an intensely religious man who also believed in witches and diabolical possession. He became swept up in supporting the veracity of accusations of witchcraft made against a local woman, a sensational episode that captured the attention of Bristol's citizens in 1762. He also corrected learned scientific treatises by his electrical friends while at the same time being drawn to Methodism, where, at one meeting house, he was able to view a new electrical machine. Although well versed in Newtonian mechanics as taught by Ferguson, Dyer distrusted aspects of the Newtonian tradition and described William Whiston, the early Newtonian, as a "Deist."[11] By his own admission Dyer was largely uninterested in business matters, and he displays no interest in industrial developments or in the merchantile world of his city, despite the fact that so many of its merchants were Dissenters or Methodists, as was Dyer himself. In him we see the assimilation of scientific knowledge by a practical man for whom religion remained the central preoccupation of his life. We can compare him with a contemporary Bristol schoolmaster, John White, also devout as his diary reveals, who gave his students "A Train of Definitions according to the Newtonian Philosophy." White apparently possessed no mystical tendencies, and one of his major intellectual interests, if we can rely on his diary, was in the new science. His Newtonian definitions closely followed the outline of any number of the lecture courses described earlier, and in his classes he dwelt at length on gravity, pulleys and levers, the laws of motion, hydrostatics, and electricity, in that order.[12] By the mid-eighteenth century such classes were commonplace not only in Bristol's grammar school (i.e., secondary or high school) but also in a variety of technical and mathematical schools intended to equip boys for practical careers.

Indeed contemporary clerical opponents of Newtonian science, and Bristol and its environs were particularly rich in one such group called the Hutchinsonians,[13] believed by 1774 that the new science, which they regarded as a threat to Christianity, had permeated the minds of the landed gentry in deepest Somerset.[14] In that year among the books proposed for purchase by the Bristol Library were Benjamin Franklin's *Letters on Electricity,* the *Philosophical Transactions* of the Royal Society, s'Gravesande's and

Voltaire's works on Newtonian philosophy, and a materialistic essay from the French Enlightenment, Helvetius's *De l'esprit* (1748).[15]

The Problem of the Bristol Port

One wonders if Richard Bright, or any of his fellow merchants in the Society of Merchant Venturers, requested those books, or had been taught by White, or had to debate with John Hutchinson.[16] If they had been taught by that Newtonian schoolmaster, they learned well—and none too soon. In the last decades of the century they were forced to bring their scientific knowledge to bear on a complex engineering problem centering on Bristol harbor, a matter critical to the maintenance of their prosperity and ultimately to the commercial as well as industrial future of the city. As we have seen a multiplicity of explanations of natural phenomena existed simultaneously amid Bristol's population—astrology, witchcraft, divine intervention, electrical principles, mechanical and Newtonian models—yet only the last would be brought to bear by engineers and merchants attempting to come to terms with the problems presented by the extreme tides of the River Avon and their effects on Bristol's harbor.

By the late 1750s the growth of commercial life and material consumption, which provided the money for transportation improvements and industrial development, began to overwhelm Bristol's harbor and rivers. The number of coastal and river craft sailing in and out of Bristol rose from an average of 900 a year in the 1750s to well over 1,700 a year in the 1770s.[17] But the exceptional tides—often over forty feet—meant that when ships in the harbor unloaded their cargoes at low tide they literally sat on the mud banks, presenting a bizarre sight described by Alexander Pope as "a long street, full of ships in the middle and houses on both sides, [looking] like a dream."[18] But the dream quickly became a nightmare when ships listed over, causing loss of the cargo or damage to the vessel, or when the absence of water at low tide made possible the spread of serious fires leaping from docks to ships or vice versa. In addition the River Avon, the main access route to the harbor, was treacherous in places, and large ships had to be towed in by row boats. In theory the corporation, that is, the town government of Bristol, was responsible for the maintenance and improvement of the river and harbor; in practice the council had delegated its responsibility to the Society of Merchant Venturers.

The society was a most elite body, composed of only the wealthiest merchants in the city and its environs, with an average annual membership throughout the century of fifty to sixty men. Among them were landed gentlemen, some with aristocratic titles; at a meeting of 1776, fifteen of the assembled possessed the title of baronet or above.[19] Great merchants and landed gentry had intermarried and secured common interests in England from at least the sixteenth century. In this Bristol society and the attempt it made to come to terms with its economic interests, and hence with the need to improve the city's transportation system, we can test the mechanical knowledge available to the mercantile and landed elites who led the Industrial Revolution in so many places.

The decision to proceed with plans for the improvement of Bristol's harbor was taken so slowly by the Society of Merchant Venturers that by the time the work was actually undertaken, in the first decade of the nineteenth century, the commercial lead had passed to Liverpool, with its excellent system of new canals giving access to the industrializing Midlands. The explanation for the delay seems ultimately to lie in the extreme wealth of the society's members, who felt no economic need to compete at that moment with their distant northern rivals and who may also have feared that improvements to the port would enhance the wealth of local small industrialists who operated outside Bristol. Not least, the "middling sorts" of citizens saw the tax levies needed for these schemes as just another instance of their having to pay out of their own pockets for the profits enjoyed by Bristol's mercantile elite.[20]

Indeed Bristol was a socially troubled city in the eighteenth century. Many of the hand workers in the city and environs had been proletarianized decades before industrialization began on a large scale in the north and in the Midlands.[21] There were riots in the 1750s in the coal mines around Bristol; and in the 1790s magistrates ordered the troops to fire on rioting citizens. In this highly commercialized city only a small proportion of the population enjoyed the surplus capital derived from the flourishing

Atlantic trade, and the extremes between rich and poor were particularly visible.

Amid these social and economic tensions the Society of Merchant Venturers took up the matter of transportation improvement. But how were mercantile gentlemen to proceed with matters as complex as tidal currents, sluices, new canals and dams, the possible installation of steam engines for drainage and the pumping of cleaner, less saline water to the city, and not least, the problem of sanitation if the water, once trapped in the harbor for the benefit of the ships, should become stagnant and polluted? Any one of these problems might have been commonplace enough. Indeed elsewhere in the 1760s canals on reasonably flat terrain were being built by engineers, such as James Brindley, who possessed no sophisticated mechanical knowledge. But taken in total, Bristol's harbor and rivers posed one of the most difficult engineering problems of the century.[22] Fortunately the rich archives of the city provide unique evidence of the extraordinary number of plans or schemes that were put forward from the 1760s onward by engineers and other natural philosophers; but more important, these records allow us to follow the gentlemen merchants as they make their way through these extremely sophisticated (for their time) mechanical discussions and, most fascinating, offer their own opinions or even corrections of engineering plans.

The society possessed sufficient familiarity with the latest mechanical techniques to seek the services first of John Smeaton and then of William Jessop, probably the best engineers of their day. But for these merchants technical matters did not end there. In our own highly specialized world, where in the first instance all scientific knowledge has become the province of highly trained scientists or technicians, such a specialist, once chosen for his skill by industry or government, would be allowed to get on with the task of designing plans and executing them, provided these activities were carefully reported and always assessed in relation to cost feasibility. In the eighteenth century cost was certainly an important factor in all of the society's discussions of engineering proposals; but so too were the engineering plans themselves. The society became the arbiter of mechanical knowledge, with rival engineers and natural philosophers, competing for approval of their plans, appearing before the subcommittee of the society concerned with these mechanical matters and never, as far as the records indicate, adopting a nonspecialist vocabulary for the society, although one was certainly used when addressing the general public and trying to convince them of the accuracy and economy of a particular plan.[23]

In 1765 John Smeaton presented the society with "Proposals for laying the Ships at the Quay of Bristol constantly afloat, and for enlarging this part of the Harbour by a new Canal through Cannon's Marsh." To get a sense of this proposal and its degree of complexity, we must read a portion of it along with the society, and so I quote from Smeaton at some length:

First: It is proposed to keep the water in the quay and new canal to the constant height of the 15 foot Mark upon the lowermost marked staff upon the quay next the Avon and by clearing away 2 or 3 feet of mud there laying to make from 17 to 18 feet of water. N.B. the 15 feet mark is about 6 feet below the top of the quay and about 4 feet below the spring tide high water mark of 24th and 25th January 1765 which, though not the largest, were never the less accounted considerable tides.

Secondly: It is proposed to dig the new canal as far as the sluices so deep as to make 18 feet water therein at the said proposed level, and to make the same at least 100 feet wide in the clear.

Thirdly: to drop the tail of the new canal into the River Avon at the bottom of Cannon's Marsh just above the Glass house.

Fourthly: to construct two separate sluices, one as near as conveniently can be to the River Avon upon the tail of the canal, the other at the distance of 400 feet from the former further within the canal, both these sluices to be furnished with two pairs of pointing gates, one pair in each sluice pointed to the land, the other to seaward. The width of the chamber of space intercepted between the two sluices to be 60 feet and the width of the sluices to be capable of taking in the largest ships that use the port which I suppose will be done by an opening of 30 feet wide.

Fifthly: The thresholds of the upper sluice to be laid at the depth of 18 feet below the constant water, that is even with the bottom of the canal but the floor of the chamber between the two sluices, as well as the threshold of the lower sluice, to be laid as low as the bottom of the river in the shallowest place below the tail of the canal.

Sixthly: These things being executed as before mentioned, the present mouth or opening of the River Froome into the Avon to be stopped up by a solid dam of earth; furnished, however, with such draw hatches, as may be necessary to assist the hatches in the gates of the sluices in discharging the freshes of the River Froome in rainy seasons; but yet as so to make a communication for all kinds of carriages from the back and quay down along side of the new canal between the same and the river.

Seventhly: The whole of the new work to be wharfed with stone, so as to form quays or such other conveniences as shall be found necessary.

Eighthly: To erect draw hatches at the new bridge at the head of the quay, capable of retaining the water behind them, when the water in the new canal and quay is let off.

Ninthly: To have the command of the hatches and to erect new ones if necessary upon the pond of Newgate Mill. N.B. By the construction above proposed, Tombs Dock and Bridewell Mill will be rendered useless, but Newgate Mill will be benefitted thereby.[24]

These proposals were accompanied by drawn plans, making visible the proposed changes, and they were followed by an explanation of how the entire system would operate. Its success depended on correct estimates of the volume and therefore the weight of the water, and the force and pressure of the tides to be admitted or closed out of the canal chambers, permitting the vessels to enter and leave them safely.

Building a canal and sluices on relatively flat terrain was not in itself extraordinary and there were time-worn procedures for doing it. What is important in these Bristol proposals is the sheer size and complexity of the problem of controlling two rivers and the tides in such a manner as to keep the harbor constantly filled with water. The cost estimated by Smeaton for the implementation of his plans was £25,000, and "at a numerous meeting of the merchants" he was unanimously thanked for his proposal. But the matter of Bristol harbor only began there.

Another mechanist, William Champion, who was a successful local industrialist in part because he was the first person to develop a chemical procedure for "making" brass (a compound of zinc), put forward his own set of complex plans for the harbor. He proposed not only a dam for the River Froome but also one for the River Avon, and he added a further proposal for erecting a steam engine "to serve the city with water at a less expense than the present great water wheel aqueducts."[25] In so doing he introduced a subject that would plague these schemes for Bristol's harbor in the years to come, and added yet another scientific question that would require expert assistance and upon which the Society would also have to pass judgment. If the water in Bristol's harbor went too low or was trapped by these new dams, the result would be stagnation and pollution from the sewers that dumped waste into the harbor, and with that would come disease. The sewer system in Bristol was already poor; these schemes for a wet harbor would, it was argued, make it worse.

In appealing for the assistance of engineers the Society opened itself to a flood of conflicting proposals from engineers, mechanists, or natural philosophers who, as one projector put it, did not "write this with any hope or intention of getting a job. I do not profess engineering, but having been many years a teacher of experimental philosophy, my experience in hydrostatics gives me full confidence in the above effects."[26] The conflicting babble of scientific tongues played into the hands of those merchants who opposed any further improvement in the harbor, who found the existing situation sufficiently comfortable as to warrant their doing nothing. The matter dragged on for many years, and then came the American Revolution. The concomitant trade embargo imposed against the colonies intervened to depress Bristol's economy and to delay the issue of the harbor into the late 1780s.

By then, however, the Society of Merchant Venturers had within its own leadership a man of the new science, Richard Bright, F.R.S., who had been taught chemistry by Priestley as well as the latest mechanical philosophy at the Dissenting academy at Warrington.[27] Bright was a merchant capitalist and a landed gentleman—he was worth £70,000 in personal and landed wealth in 1777—in other words, a member of the urban gentry and oligarchy whose Whiggery complemented his scientifically grounded faith in the necessity of progress and improvement. He made the project of promoting Bristol's harbor his own personal crusade, and he enlisted in this struggle all his natural philosophical talents and contacts, as well as his political influence. He saw more clearly than many of his contemporaries in the society, of which he was for a time secretary, that its profits hinged on the improvement of the harbor, that this was essential in order for

Bristol to compete effectively with Liverpool.[28] He sent copies of new engineering proposals to his friends in the Manchester Literary and Philosophical Society for their approval, and in so doing he demonstrates to us his identification with a scientific society at the forefront of the industrial application of science.[29] He also sought out optimistic medical opinion that would contradict the opinion of medical experts who had judged various plans for the harbor to be dangerous to the sanitation system of the city. At the end of one such favorable medical report, offered by a Dr. Falconer, F.R.S., Bright penned in his own expectation "that nothing will impede the prospect of our improvement."[30]

Once again, as in the 1760s, the society was deluged by contradictory engineering plans, but in the 1780s new factors were evident. Not only are the plans more complex and more expensive to implement, but conflicting medical evidence has also become a public issue. The professionalization of scientific knowledge is everywhere evident; as one fellow of the Royal Society commented to Bright: "I will decline giving any answers to the queries you sent relating to the proposed undertaking at Bristol as none but Physicians are proper judges of many of them and the Engineers they [the Society] have consulted are much better judges of the remainder."[31] Once again the best engineers are enlisted, only now the Society can no longer put off its decision.

In the course of the ensuing debate within the society its secretary, now another local merchant, Jeremiah Osborne, was instructed to query the very theoretical principles on which Smeaton's hand-picked successor, William Jessop, based his engineering plans. Obviously, the discussions within the society had ranged to the very philosophy of nature that lay at the basis of these complex proposals. Jessop's response was a short lecture on mechanics, just the sort of basic scientific information that was disseminated up and down the country by the itinerant lecturers; only Jessop confesses that he has forgotten some of the fine points of the new mechanics:

> To make you completely acquainted with the Principles on which the Calculations are founded, respecting the discharge of Water over cascades or through orifices would take me much time and some study; for having in the earlier part of my time endeavoured to make

myself acquainted with these Principles, and having been once satisfied with the result, I have, as most practical men do, discharged
my memory in some measure from the Theory, and contented myself
with referring to certain practical rules, which have been deduced
therefrom, and corrected by experience and observation. But I can
in a few *Weeks* inform you in the general principle on which these
calculations are grounded. It has been found by experiment that a
heavy body falling from rest will descend about 16 feet in a second
of time; and that the velocity acquired at the end of that second is
such, as proceeding on at the same rate, or without any acceleration,
would carry it on in equal time through a sphere of double the height
which it fell from, or 32 feet in a second; that bodies falling from
different height acquiring velocities in proportion to the square root
of these heights. And that water in going through orifices runs with
the same velocity as would be acquired by a heavy body in falling
through a space equal to the height of the waters surface above the
virtual center of the orifice. So that while a height of 16 feet would
produce a velocity of 32 feet in a second, a height of 9 feet would
produce a velocity of 24 feet in a second or as 4 the square root of
16 is to 3 the square root of 9. But as this is the greatest possible
velocity that can be acquired; it is found in practice to vary from this
rule, conformable to a variety of circumstances, such as the shape of
the orifice, the manner in which the water is introduced into the
mouth of it, the friction in passing through it, etc. and so that while
in some cases it might not discharge about 2/3 of the full quantity,
in others, it fluctuates between that and the full quantity, in degrees
which only experience and nice observation can nearly ascertain. So
in flushing over cascades it is found that the velocity is something
less than would be due to a height equal to half the thickness of the
sheet of water, for instance, in a sheet of 18 inches thick, the velocity
would be such as would be produced from falling a height of about
eight inches or about a fifth part of what would be due to 16 feet.
If these hints will threw light upon your enquiries it will give pleasure
to your most obedient servant, W. Jessop.[32]

The basic principles of the mechanical philosophy, coupled
with observation and experimentation—as described by Jessop—
along with the professionalization of mechanical science and its
application by practicing engineers, had come to be accepted by
entrepreneurs and merchants alike. These are the elements that
went into making what we may reasonably describe as the industrial mind. Armed with this mechanical understanding of nature,

and willing to give credibility to the superior knowledge claimed by professional mechanists, merchants, entrepreneurs, and industrialists could, and did, make decisions that formed an essential part of the history of the early Industrial Revolution.

By the 1790s the merchants of Bristol were finding it necessary to comprehend engineering plans that described nearly £200,000 worth of alterations and land purchases which were now needed to improve the harbor (see illustration). Jessop's technical drawings were accompanied by descriptions such as the following:

> G. - shows the falling doors turning upon the axes H, which are held in their upright position by cast iron weights, I, so balanced that whenever the water rises above its common height the pressure against the doors will increase, the weights will be raised, and the doors will lie flat, leaving nothing projecting, or liable to derangement by ice, or other floating substance. Upon the subsiding of the water, the weights will again overcome the pressure on the doors, and raise them to their upright position.
>
> AB. - is a cylinder of 5 feet diameter open at the bottom or base on which it stands, closed at the top, and perforated by four large apertures at the sides, or in the circumference.
>
> C. - The cylinder, C, is suspended to a beam which moves on a Center D, and at the other end of this beam is suspended a cast iron bucket E, which moves up and down in a well. When the water rises above its common height, it will flow through a pipe F, and fill the Bucket, causing it to preponderate and raise the cylinder C. . . . I must observe that as the pressure of the water counteracts itself on all sides of the cylinder C, it will move under any head of water without much friction. (I believe this cylinder was the invention of the ingenious Mr Westgarth, and applied, on a small scale to an engine for raising water from mines [i.e., a steam engine] and has been used by Mr Smeaton for a similar purpose).[33]

We might be tempted to imagine that in the face of all this technical verbiage the merchants who favored the harbor, led by Bright, simply consigned themselves into the hands of Jessop, the most famous and accomplished engineer they could find, a man whose prestige could sway public opinion and stand up well under parliamentary cross-examination. For in the first instance, Parliament would have to legislate on the basis of the plans

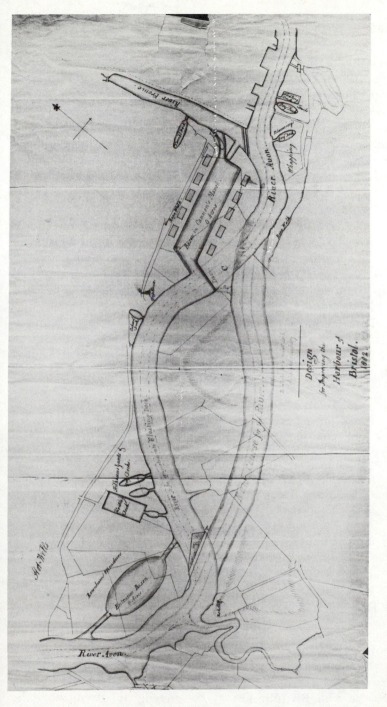

From the Bristol Record Office, Bright MSS, MS 1116(i)e. (*Courtesy of the City of Bristol*)

235

presented to it by the Society; an act was required whenever private land was to be requisitioned and purchased, or businesses were threatened by the diverting of local water supplies, or money was to be raised by the selling of shares to the public. But the letter and minutes of the society, as well as Bright's private notebooks, show the merchants themselves discussing engineering plans in some considerable detail, sitting through complex sessions with engineers—discussing water levels in the harbor, or weighing ecological objections based on health considerations and sewage control, or deciding the merits of placing steam engines on various docks.[34] The merchants became the final arbiters of scientific knowledge and its application. They were capable of assuming this role because they had learned enough about the mechanical philosophy, through reading or lectures or personal experimentation, that they were able to apply it at this level of sophistication.

In 1792 Jeremiah Osborne had to arrange with a London engraver for the printing of Jessop's hand-drawn plans. But he advised the engraver to hold off drawing one section and noted his disagreement, based on direct observation, with Jessop's figures:

> One of the observations I have made on the sections and which I wish cleared up before that part of the Plate is begun to be engraved, is that the 15 foot level on the quay appears to be rather above the 14 foot mark on Hillhouse Dock gauge in the section of the Avon; but I believe that the bottom of that gauge should stand six feet higher than it does in the drawing so that 13'6" on it should answer to the 15' mark.[35]

As it turned out Osborne's confidence in his knowledge was well founded; he had caught an error in the engineering plans of William Jessop.[36] The society also took the advice of one of the medical experts who wanted two sluices on each side of a dam, to prevent flooding.[37]

Bristol in the Nineteenth Century

Finally, in 1804, work on the floating harbor began; the improvers had won out. The problems of class resentment so commonplace in Bristol's history continued, and predictably the company formed to execute Jessop's plans became an object of resentment. It alienated many local citizens by its secretive and high-handed manner. One irate citizen warned the directors "you may chance to get a ducking, in that stagnant lake, to which your own enlightened conceptions have given birth."[38] Bristol's lead in industrialization, based on the iron foundry business of Abraham Darby and its local brass and glass industries, nevertheless gave way to its northern rivals.[39] Its merchants had made their decisions on the basis of technical knowledge, but they had done so too slowly. Industrial dynamism now came in particular from Birmingham, where the application of steam power to the glass and brass industries was coupled with a better canal system, and hence glassware and "toys"—that is, small brass objects—could be made more cheaply and transported more efficiently. Yet it should not be imagined that Bristol's elite lost out entirely; the city continued as a banking and commercial center of importance throughout the nineteenth century. By 1825 it possessed its own literary and philosophical society, where lectures on natural philosophy, very similar to those that abounded earlier in other eighteenth-century provincial cities, were commonplace and offered the latest scientific knowledge on electricity and magnetism.[40]

THE POLITICS OF MECHANICAL APPLICATION

As we can now see, a multitude of factors went into successful industrial and commercial decision making: the ability to comprehend increasingly complex technical knowledge through a mastery of the mechanical philosophy, the presence of entrepreneurs willing and able to push a particular project through Parliament, sufficient surplus capital from large and small investors to be invested in shareholding companies, and never least, the availability of laborers to dig the canals or feed the furnaces. Scientific knowledge was only part of the story in Bristol and elsewhere, but it was a vital part.

The industrial application of scientific knowledge constitutes historically the single most important use to which Western science has ever been put, and this occurred first in England in the second half of the eighteenth century. Obviously there were many moments in the early Industrial Revolution when scientific knowledge, particularly in the area of transportation, mattered not at all.[41] Indeed the basic techniques of simple canal building and water control had been used in ancient China as well as in seventeenth-century France and Holland. But in the late eighteenth century in England these techniques first became commonplace and eventually mechanically sophisticated beyond recognition.

In addition, as early as the 1760s reformers seized on canal building as the fulfillment of public interest at the expense of "self-interested motives and local views."[42] The pioneers of early canal development, without which the coal needed by the industrial centers would never have been made available cheaply or in sufficient quantity, saw themselves as reformers, in opposition to monopoly interests. As we have seen in the example of John Smeaton's anger when engineers were not properly rewarded (pp. 162–163), these projectors and developers, whether involved in transportation or industry, could employ the political language of the opposition when their interests were thwarted; but they largely identified with the Whig oligarchy, and they sought out and, most important, received its support. This point needs to be stressed as we survey the uses to which science was put in this period; namely that its successful application required support from the landed classes that controlled Parliament. From the 1760's into the 1790s all evidence suggests that the Whigs* in particular were identified with industrial innovation and canal building.[43] Once again we find secular-minded elites drawn to science as a way of increasing their wealth and power; only in late eighteenth-century England, political stability and centralization permitted this application to become national in character, with consequences that would soon become global.

By the 1790s we can see how technically sophisticated the revolution in transportation had become by surveying the records of canal companies from the period; and once again, those records illustrate the depth of mechanical knowledge to be found among gentlemen of land, industry, and commerce. The example of the Bristol merchants can be multiplied across the country, although few towns, counties, or cities, to be sure, faced the

*See glossary of terms.

complexity of engineering problems presented by Bristol's harbor. Sometimes the records also clearly indicate that technical knowledge which could have been used to great advantage was simply absent, and the results were often disastrous, as lives were lost and money wasted on canal projects that were badly designed.[44]

As the mania for canal building swept the country in the last decade of the eighteenth century, hundreds of canal companies were formed, and engineers like Jessop grew to be wealthy men as their services were widely sought. In the north of England the same engineers who designed the canals frequently invested in their companies or became industrialists in their own right.[45] Indeed there the linkage between industrial development and canal engineering was perceived almost immediately; but in case Parliament or local interest groups needed reminding of the necessity for new systems of transportation, natural philosophers like Erasmus Darwin stood ready to admonish them and to urge on the projectors.[46] Not every industrialist of this early period supported canal building or bothered to understand the mechanical principles used by canal engineers in making their plans or by mechanists such as James Watt and Matthew Boulton in designing and installing their steam engines. Even renowned industrialists—Richard Arkwright, for example, famed in part for his self-taught mechanical ability—opposed improvements that might threaten their profits.[47] And, of course, there was competition from older monopolies where earlier improvements had served to entrench their wealth and position. But where a neat fit existed between profit and improvement, we once again see merchants, landed gentlemen, industrialists, engineers, and natural philosophers allied in common interest, abetted by the mechanical knowledge they held in common.

The landed capitalists and factory owners who hired an engineer, as for example Philip Gell, and the promoters of the Cromford Canal in Derbyshire, who hired William Jessop in 1788 and asked him to draw up plans for it, sought always to get the best mechanical knowledge that money could buy. They knew that whomever they hired would have to go before parliamentary committees, where lords could be found who were "for teaching anybody [their business,] the Bishops', religion; the Chair, the law; and now Jessop as an engineer."[48] When a mere "teacher of

Mathematics and Philosophy," of whom they had never heard, was brought in to testify against the canal and managed to get his calculations close to Jessop's, the promoters were genuinely surprised.[49] They had come to accept the professionalization of scientific knowledge of a mechanical sort and to rely solely on engineers, preferably famous ones, if they could be found. The promoters sat through parliamentary cross-examinations of these experts and followed their estimates of the weight of water lost through the diversion of river water into a canal, and in some cases they appear to have understood more about the mechanics involved than did the lawyers who were doing the questioning.[50]

That knowledge was necessary for lobbyists if their case was to be won against those who felt they stood to lose water power for their factories as a result of diversion by the canal and who brought in their own mechanical experts to argue against the proposed canal bill. At moments in those hearings less knowledgeable witnesses on both sides damaged their client's case. We can watch the self-satisfied response of canal promoters who happened to possess greater mechanical knowledge and used it more effectively before these parliamentary committees. The promoters and engineers alike recognized the necessity of making actual "experiments" in order to be able to present the most accurate information before Parliament.[51] On those occasions lords on the committee could be seen "constantly taking notes and making good observations and asking really very pertinent questions."[52] Others, however, admitted that they simply "did not understand it." Mechanical knowledge among the English aristocracy was widespread but not universal.

The best guide to the depth of that knowledge and its application lies in the parliamentary records of those committee hearings. These document moments of decision making in the early Industrial Revolution that reveal the key role played by central government, particularly in the transportation side of that revolution. Without the canals, harbors, and turnpikes approved by parliamentary bills, that revolution would have been stillborn. Many factors went into those parliamentary decisions: political pressure brought by local interests, the reputation of the engineers, and outright bribery; but not least was the mechanical knowledge of the committee members and their enlightened faith in the value of improvement.

In the hearings for the Cromford Canal, for example, the committee pressed the matter of the effect of the lost water pressure on the profits of some factory owners and the concomitant danger of increasing unemployment in the district. Time pieces were fitted on water wheels in order to supply that evidence, while expert witnesses were called in to testify on the relationship between water pressure and the power of turning wheels, to explicate as one natural philosopher put it "by known hydrostatick Principles agreed to by all authors,"[53] or to give evidence based on conversations "with many who are scientific, and I have read most books upon the subject."[54]

The questions and answers, the flow of discussion on these occasions tells us much about the use of mechanical knowledge in this period and, most important, about the ease with which it was assimilated and used. Witness this cross-examination of one Richard Roc, "a surveyor and teacher of mathematics," questioned by the House of Lords committee:

—You are a surveyor and teacher of mathematics.

Yes.

—Suppose two shutters of a Mill of 4 feet each are elevated 17″ with 4 feet over them. What quantity of water flows in a minute?

278 Tons per minute.

—That is when the water is 4 feet high from the bed of the river?

Yes.

—Do you speak from observation or calculation?

I calculated it from the Dimensions given by Mr Snape.

—How do you ascertain it?

By known Hydrostatick Principles. . . .

—Then can you say that with a given velocity a certain quantity of water flows in a minute?

Yes.

—Do you make that Calculation upon a supposition that the water is not retarded by water contiguous to the wheel?

Certainly.

—Then in point of fact, supposing the water to be retarded, would not the quantity be less?

Yes, very much less.

—Cross-Examination: Have you ever measured the River Derwent at
Cromford Bridge?
Yes.

The questions proceeded from general theoretical principles to
Cromford's proposed canal in particular and ended, once again,
on the theoretical:

—What is the rule by which you stake your calculations?
From the height a body falls in a second of time, it is said to fall
16', 1" in a second, and then acquires a velocity which will take it
through twice that distance, I then proportion it by the square root
of the height.[55]

When the lords turned their attention to the vexed question
of the interests of factory owners dependent on water power for
their profit and afraid that the canal might decrease that power,
engineers in favor of the project presented elaborate mechanical
arguments to refute those objections.[56] There are poignant moments in these deliberations—for example, when a mill foreman
whose livelihood had already been adversely affected by the
canal, is asked:

—Could you increase the power of these wheels still more if you
tried?

His answer reveals that he simply does not understand the mechanical principles that have already been applied to the detri-
ment of his water supply:

Upon my word, I do not know, for the power of a wheel is what I do
not understand.[57]

Doubtless there were many mill owners who did not understand the power of a wheel in mechanical terms. Indeed one of
the myths about the Industrial Revolution is that few, if any, of
its progenitors understood the science most commonly disseminated throughout the eighteenth century. We know of a few such
early industrialists who apparently possessed no theoretical

knowledge, yet it is also possible to find historical evidence that significantly contradicts the myth.

The Steam Engine

If we go to the lead mines of Derbyshire, to the industrial center of economic development in the late eighteenth century, there too we find evidence of mechanical knowledge applied by mine owners who possessed no known formal, academic scientific education. The business of deciding to install a steam engine was tricky; bankruptcy quickly followed if the wrong engine was installed in the wrong place. In 1794 a steel firm failed immediately after installing an engine; it was "too heavy [too expensive] a concern."[58] Yet it was obvious from as early as the 1720s—as Desaguliers, Martin Clare, and the scientific lecturers pointed out —that the steam engine had enormous potential, particularly in mining, where its power could be used to extract water from the subterranean tunnels, which were always prone to flooding. By the 1770s innovations in coal mining, at least in the Derbyshire area, were being introduced by men with experience in mining who also now possessed geological knowledge as well as an understanding of the Newcomen engine and the significant improvements made on it by Watt.[59] These were lead merchants, such as Benjamin Wyatt and John Barker, who had a wide knowledge of their industry and of the principles of commercial life, but also, at least in some cases, of the technical and theoretical aspects of the mechanical philosophy.

The historical literature about the early Industrial Revolution tends to describe the use of steam power in production as if its application had been an automatic process. The benefits of the engine were perceived immediately; if capital could be found by owners they simply called in engineers who installed the necessary engine. The engineers knew mechanical science, the owners did not—or so the argument goes. They made the decision to install solely on the basis of economic considerations, cost of fuel, labor, and so forth, and their own relationship to the machine and its power was largely passive and inarticulate. But many of these owners were smarter than historians have allowed them to be. They knew that there were many variables and so much at stake that it behooved them to think well on these machines, to understand what they could and could not do. Lead mine owners recorded their anxiety about what a steam engine could do: "How this may answer no one can say; so much depending on accidents."[60]

Where we can find evidence of the process of consultation between owners and engineers, there we find the "experts" speaking in considerable technical detail to their employers, complete with mechanical drawings.[61] Sometimes encouragement to proceed with installation came from Sir Joseph Banks himself.[62] As president of the Royal Society he had an interest in the application of the mechanical science that so many of its Fellows had been instrumental in disseminating. He also had investments in mineral veins from which he sought to profit. Aside from taking advice from Banks or others, mine owners went about the countryside observing steam engines at work and then told the engineers what they wanted. They sought estimates and again, like those Bristol merchants, had to choose among designs.[63] They had to comprehend the technical data laid before them and, as did William Wyatt who inherited Benjamin Wyatt's successful lead mining business, they knew in the end what they wanted:

I have through the hand of mr Snyed received your estimate for a steam engine of sixty horse power but before any further steps are taken I will thank you to furnish me with the following particulars, viz., the diameter of the cylinder, the construction and size of the boilers, the length of the beam, the weight of the ply wheel, the number of strokes to work per minute, the quantity of water lifted 240 yards deep at each stroke, the diameter of the working panels and the quantity of coal which will be consumed in every 24 hours. Perhaps we can use a plunger to advantage as our pitt will be a great depth. Please write me by return of post as I am anxious to determine about an engine soon as possible. Are there any very superior pumping engines to be seen at work in your county, if so I should like to see them—I mean engines that do a deal of work with a little fuel.[64]

Levers, beams, pulleys, and weights—the stuff of those experiments by which the new mechanical philosophy was illustrated in books and lectures—when combined with the profit motive, cheap fuel, and access to transportation for raw materials as well as consumer goods—transformed the means of production first in England, then in western Europe. When we ask ourselves why this happened first in England, we must remember the English Revolution and the relationship it forged between its landed and commercial beneficiaries and the new science (see Chapter 3). Other Western elites aided by progressive intellectuals in various Continental countries aspired, as we have seen (Chapter 6), to the industrial application of mechanical science. But political factors—perhaps more critical than the presence of coal or surplus capital or labor—would delay that process in France and The Netherlands until the nineteenth century. By then the English industrial model existed in reality, no longer solely in the dreams of the natural philosophers. It was in part the consequence of a series of discrete decisions made by entrepreneurs who aspired to scientific knowledge because self-interest and improvement demanded that they have it.

NOTES

1. Peter Mathias, *The First Industrial Revolution: An Economic History of Britain, 1700–1914* (London: Methuen, 1983), pp. 128–29; or see E. A. Wrigley, "The Supply of Raw Materials in the Industrial Revolution," *Economic History Review*, vol. 15 (1962), p. 4: For a useful corrective see D. S. L. Cardwell, *The Organisation of Science in England* (London: Heinemann, 1972), pp. 13–18; Alan Smith, "Steam and the City: The Committee of Proprietors of the Invention for Raising Water by Fire," *Transactions of the Newcomen Society,* vol. 49 (1977–1978), pp. 5–18, on the Royal Society and the steam engine. For one of the first cogently argued attacks on the view represented by Wrigley, see A. E. Musson and E. Robinson, "Science and Industry in the Late Eighteenth Century," *Economic History Review,* 2nd ser., vol. 13 (1960–1961), pp. 222–244, especially pp. 241–242 for further evidence of scientific lecturing in Bristol and Sheffield.
2. I am not seeking here to reintroduce a retrograde "idealism" into historical analysis, rather to warn social historians of the distorting effect their discipline can have. Cf. R. S. Neale, *Class in English History* (Oxford: Blackwell, 1981), pp. 140–145.
3. Karl Marx, *German Ideology,* ed. R. Pascal (New York: International Publishers, 1963), pp. 36–37: "The class that has the means of material production at its disposal has control at the same time over the means of mental production, so that thereby, generally speaking, the ideas of those who lack the means of mental production are subject to it."

4. William Chapman, *Address to the Subscribers to the Canal from Carlisle to Fisher's Cross* (Newcastle, 1823), pp. 2–3, 7. This essay was written as a result of a series of breakdowns in relations between an engineer of the next generation and the canal company. Emphasis in the quotation from Smeaton was added by Chapman.
5. Chapman, *Address to the Subscribers,* p. 2.
6. Anthony Burton, *The Canal Builders* (London: David and Charles, 1981), pp. 157–158; and R. W. Malcolmson, *Life and Labour in England 1700–1780,* (London: Hutchinson, 1981), pp. 83–93.
7. For a general discussion of Bristol in this period, see B. D. G. Little, *The City and County of Bristol: A Study in Atlantic Civilization* (London: Werner Laurie, 1954).
8. See Thomas A. Ashton, *Iron and Steel in the Industrial Revolution* (Manchester: Manchester University Press, 1963), pp. 21–30, 41–42; Brian Bracegirdle, *The Darbys and the Ironbridge Gorge* (London: David and Charles, 1974); and Isabel Grubb, *Quakerism and Industry Before 1800* (London: Williams and Norgate, 1930), pp. 50–51, 151–55.
9. Bristol Central Library, MS 20095, "Diary of William Dyer," vol. 1, 1760, fol. 116. For an outline of the lectures Ferguson gave up and down the country, see James Ferguson, F.R.S., *Lectures on Select Subjects in Mechanics, Hydrostatics, Hydraulics,* 6th ed. (London, 1784), an overtly Newtonian course, very similar to those discussed in the previous chapter. I am grateful to Jonathan Barry for information on Dyer.
10. "Diary of William Dyer," vol. 1, 1760, fol. 111, for this description of her; 1763, fol. 116, for the evening in question.
11. Ibid., fol. 126.
12. Bristol Record Office, White MS, no. 08158, fols. 73–81.
13. See Roy Porter, "Alexander Catcott: Glory and Geology," *British Journal for the History of Science,* 1977.
14. Bristol Central Library, MSB 26063, correspondence of Rev. A. S. Catcott and A. Catcott, letter of June 23, 1774, to A. Catcott.
15. Bristol Central Library, Bristol Library MSS, "Books proposed 1774," written in a variety of hands. For later developments, see Michael Neve, "Science in a Commercial City: Bristol 1820–60," in Ian Inkster and Jack Morrell, eds., *Metropolis and Province: Science in British Culture 1780–1850* (London: Hutchinson, 1983), pp. 179–204.
16. Of the 155 pupils at Bristol Grammar School from 1710 to 1717, 53 became merchants and mariners. For the considerable education given to the sons of wealthier merchants, see W. Minchinton, "The Merchants of Bristol in the Eighteenth Century," *Sociétés et groupes sociaux en Aquitaine et en Angleterre* (Bordeaux: Federation historique du Sud-Ouest, 1979), pp. 190–191.
17. Alan F. Williams, "Bristol Port Plans and Improvement Schemes of the 18th Century," *Transactions of the Bristol and Gloucestershire Archaeological Society,* vol. 81 (1962), p. 144.

18. Alexander Pope, *Letters to Martha Blount,* 1732, quoted in Williams, "Bristol Port Plans," p. 142.

19. For a general history of this body, with an excellent chapter pertaining to the river and harbor problems, see Patrick McGrath, *The Merchant Venturers of Bristol* (Bristol: Society of Merchant Venturers of the City of Bristol, 1975), especially pp. 150–153; and for the meeting records, see the Society of Merchant Venturers, Clifton, Bristol, Merchants' Hall Book of Proceedings, records for May 1776.

20. Williams, "Bristol Port Plans," p. 178.

21. See Nicholas Rogers, "The Urban Opposition to Whig Oligarchy, 1720–60," in Margaret C. Jacob and James R. Jacob, eds., *The Origins of Anglo-American Radicalism* (London and Boston: George Allen and Unwin, 1984), pp. 138, 142–145.

22. Williams, "Bristol Port Plans," p. 148.

23. See, for example, *Observations on the Dangers and Inconveniences Likely to attend the execution of the proposed scheme of building a Dam across the River Avon* (Bristol, 1791).

24. Bristol Record Office, Proposal of 1765, MSS of Richard Bright.

25. Williams, "Bristol Port Plans," p. 147.

26. Bristol Record Office, MS 111689(3), proposal from A. Walker, 1791.

27. Pamela Bright, *Dr Richard Bright 1789–1858* (London: The Bodley Head, 1983), pp. 13–18, on this Bright, the father of her subject. See also Royal Society, B.L.A. b. fols. 325–229.

28. Bristol Record Office, Bright MSS, 11168(3), a long list in Bright's possession that estimates the number of ships using Bristol harbor, with direct comparisons to Liverpool.

29. Bristol Record Office, Bright MSS, 11168(3), letter of November 16, 1791, Thomas Percival to Richard Bright. See Arnold Thackray, "Natural Knowledge in Cultural Context: The Manchester Model," *American Historical Review,* vol. 79, no. 3 (June 1974), pp. 672–709.

30. Bristol Record Office, Bright Mss, MS 11168(3) "opinion tendered by Dr Falconer." Bright did profess his deep concern that no "injury should arise to health," see R.S. B.L.A. b. fol. 327.

31. Bristol Record Office, Bright MSS, 11168, Henry Cavendish to Richard Bright.

32. Bristol Record Office, Bright MSS, 11168(3), November 15, 1790.

33. Bristol Record Office, Bright MSS, 11168(1)e. The plan was first submitted on February 25, 1790.

34. Society of Merchant Venturers, Clifton, Bristol, MS Letter Book 1781–1816, for example, entry for May 20, 1792, the society to Mr James Allen, on his architectural plans not to be preferred to what has been submitted; H.B. microfilm 4, December 6, 1786, a meeting where a variety of engineers appeared and presented their ideas; MS Letter Book, August 15, 1815, to William Jessop: "Your plan of the proposed Crane has been submitted to the Society. . . . Upon exam-

ining it with that of Messrs Stewart and Ramsden the Radius described by your Crane does not appear to be equal to theirs. The Arm of the Crane does not reach so far out by two feet and taking a perpendicular or plomb line from any given point of the Brace C to the level of the Wharf there is a considerable difference in the height." See also Bristol Record Office, Bright MSS, 11168(66–68), Bright's notebooks.

35. Society of Merchant Venturers, MS Letter Book, entry for July 17, 1792, to Mr Faden, engraver, St. Martin's Lane; see also letter dated August 18, 1815, to Jessop, from which it is clear that the society's committee has once again altered his plans.

36. Ibid., fol. 206, 1792.

37. Ibid., Jessop to Osborne, January 11, 1793; for a comparison of the complexity of such plans versus those available a hundred years earlier, see Bristol Central Library, Southwell MS, undated handbill at end of the volume from the 1690s.

38. *Felix Farley's Bristol Journal,* March 21, 1807, quoted in R. A. Buchanan, "The Construction of the Floating Harbour in Bristol: 1804–1809," *Trans. BGAS,* vol. 83 (1969), p. 199.

39. Little, *Bristol,* p. 167.

40. Bristol Central Library, MSS of the Bristol Library and Philosophical Institution, 1825. Cf. Charles H. Cave, *A History of Banking in Bristol from 1750 to 1899* (Bristol, 1899).

41. For a good description of the earliest partnership in canal building, which involved James Brindley, a mechanic of little or no scientific training, and a landed aristocrat, the duke of Bridgwater, see Francis Henry Egerton, *The First Part of a Letter to the Parisians, and, the French Nation, upon inland Navigation* (Paris, 1818); for James Brindley's orderly mind, see his diaries, 1759 to 1763, Central Library, Birmingham.

42. *The History of Inland Navigations. Particularly those of the Duke of Bridgewater in Lancashire and Cheshire* (London, 1766), p. 34.

43. Anthony Burton, *The Canal Builders* (London: David and Charles, 1981), p. 50; see also Derbyshire Record Office, D258/50/13/p, March 19, 1789, on canvassing Bishop Llandarff to support a canal bill, "He is a Liberal, though a Bishop." For a discussion of some of the complexities of this Whig commercialism, see J. G. A. Pocock, "Radical Criticisms of the Whig Order in the Age Between Revolutions," in Margaret C. Jacob and James R. Jacob, *The Origins of Anglo-American Radicalism* (London and Boston: Allen and Unwin, 1984), pp. 42–43. On the social composition of the early Industrial Revolution, see Harold Perkin, *The Origins of Modern English Society 1780–1880* (London: Routledge and Kegan Paul, 1969), pp. 67–68. See also Peter Buck, "People Who Counted: Political Arithmetic in the Eighteenth Century," *Isis,* vol. 73, no. 266 (1982), p. 32, on court Whigs favoring a national census in 1753.

44. See R. B. Schofield, "The Construction of the Huddersfield Narrow

Canal 1794–1811: With Particular Reference to Standedge Tunnel," *Transactions of the Newcomen Society,* vol. 53 (1981–1982), pp. 17–38.

45. See Philip Riden, *The Butterley Company, 1790–1830: A Derbyshire Ironworks in the Industrial Revolution* (Chesterfield, 1973), p. 3 ff., for Benjamin Outram.

46. See, for example, Derbyshire Record Office, D258/50/14 w, E. Darwin to P. Gell, April 22, 1789.

47. R. B. Schofield, "The Promotion of the Cromford Canal Act of 1789: A Study in Canal Engineering," *Bulletin of the John Rylands University Library of Manchester,* vol. 64 (1982), pp. 246–247. Cf. R. S. Fitton and A. D. Wadsworth, *The Strutts and the Arkwrights 1758–1830* (Manchester: Manchester University Press, 1958), pp. 62, 80.

48. Derbyshire Record Office, D258/50/14 y, to Philip Gell from his brother in London, July 7, n.a.

49. Derbyshire Record Office, D258/50/14 ta.

50. Schofield, "Promotion of the Cromford Canal Act," p. 268.

51. Derbyshire Record Office, D258/50/14 v, B. Outram to P. Gell. Cf. Schofield, "Promotion of the Cromford Canal Act," p. 274.

52. Schofield, "Promotion of the Cromford Canal Act," p. 270, quoting a letter from John Gell to Philip Gell. There is no evidence that committee members were chosen for their particular expertise; see O. Cyprian Williams, *The Historical Development of Private Bill Procedure and Standing Orders in the House of Commons* (London: HMSO, 1948), vol. 1, pp. 41–46.

53. House of Lords Record Office, Main Papers, H.L. May 26, 1789, et seq.

54. House of Lords Record Office, Main Papers, May 24, 1791, evidence on Birmingham Canal Bill.

55. House of Lords Record Office, Main Papers, May 26, 1789, Cromford Canal.

56. House of Lords Record Office, Main Papers, May 19 and 20, 1809, Kennet and Avon Canal Bill, examination of John Rennie, Esq.

57. House of Lords Record Office, Main Papers, May 19, 1809, Kennet and Avon Canal Bill. This is a bill to permit the raising of more money for a canal that is partially completed.

58. T. S. Ashton, *An Eighteenth Century Industrialist: Peter Stubs of Warrington 1756–1806* (Manchester: Manchester University Press, 1939), p. 41.

59. James H. Rieuwerts, "A Technological History of Drainage of the Derbyshire Lead Mines," (Ph.D. diss., University of Leicester, 1981), pp. 145–149. Cf. Roy Porter, *The Making of Geology* (Cambridge: Cambridge University Press, 1976).

60. Sheffield City Library, Bagshawe Collection, MS 494, John Barker's Letter Book, 1765–1811, entry for September 30, 1794, on a mine subject to a great deal of flooding.

61. Derbyshire Record Office, 503/D103, William Jessop to Mr God-

win, Butterley Ironworks, September 9, 1815, and December 14, 1815.

62. Sheffield City Library, Bagshawe Collection, C. 654(1–116), letter of William Milner to George Barker on steam engine with the approval of Sir Joseph Banks, September 21, 1807. Cf. Lynn Willies, "The Barker Family and the Eighteenth Century Lead Business," *Derbyshire Archaeological Journal,* vol. 93 (1973), p. 68, on Wyatt taking over the failing business of the Barkers and revitalizing it.

63. Sheffield City Library, Bagshawe Collection, C. 587/(30), fol. 1, estimate with technical description of engine, from R. Smith to W. Wyatt, December 9, 1836; fol 3, W. Sneyd to W. Wyatt for a 60 horsepower engine; fol. 8, another estimate with details. The cost involved is between £2,000 and £3,000; see February 9, 1837 for sums.

64. Sheffield City Library, Bagshawe Collection, MS 587(30), fol. 4, William Wyatt to Mr Cope, Bakewell, January 31, 1837. Cf. N. Kirkham, "Steam Engines in Derbyshire Lead Mines," *Transactions of the Newcomen Society,* vol. 38 (1965–1966), pp. 72–73, 76–77, on Wyatt as an innovator.

EPILOGUE

The integration of scientific knowledge into our culture may be said to be one of its most signal characteristics. The mechanical model of nature has permeated our thought processes. But to that benign statement must be added another description of our reality in the late twentieth century. The technological by-product of nuclear science in the hands of Western and a few non-Western powers, the nuclear bomb, can now be used to destroy our own culture, not to mention many other of the world's cultures. At this moment in human history historians concerned with Western science are being asked increasingly to make ethical or normative statements about the historical meaning of scientific inquiry—in effect, about how we got to where we now are. It is also true that in the postwar era a new history of science in our culture is being written by a generation of historians capable of standing back, as it were, from our science largely because the once unassailable claim that its progress equals human progress can no longer be assumed. Here I wish to speak as one of those citizen historians, bearing in mind that nothing in our highly specialized craft prepares us to be wise, when we are only at best somewhat learned.

The history here described reveals that Western science at its foundations, as promoted by its most brilliant as well as its most ordinary exponents, never questioned the usefulness of scientific knowledge for warmaking. I know of no text from the early modern period which suggests that the scientist should withhold his knowledge from any government, at any time, but especially in the process of preparing for warmaking. Indeed most texts that recommend science also propose its usefulness in improving the state's capacity to wage war more effectively, to destroy more efficiently. Early modern scientists identified with that small literate segment of the elite to whom power was presumed to be a natural prerogative. Consequently there is no long historical lineage for the position—which it may be argued must now be taken —that certain types of scientific knowledge have become inherently so dangerous to our survival that they must no longer be made available to any government, that ethical scientists of *all* political and ideological persuasions must now regard a colleague who continues to build nuclear bombs as no longer a member of the scientific or technological community. There is little in the history of discourse about science to make the construction of such an inhibition on learning and its application an easy one to achieve.

Similarly the concern increasingly expressed about the exclusivity of science, that it still remains largely the domain of white Western males, should be seen as one rooted in historical reality. Not only scientific discovery but, more important, access to its application occurred at a time of minimal literacy for the majority of Western women. Both of those exclusions were noted by early modern male commentators[1]—seldom with disapproval —although by the middle of the eighteenth century significant efforts were made by some of them to correct that deficiency among literate women. However exclusionary for women early modern scientific discourse may have been, it should be remembered that the religious and theological realm—with the exception of a few sectarian movements—totally excluded women from intellectual leadership. The domination of nature as the early modern goal of scientific inquiry was also expressed in gender-based language that made "her" a fit object for subjugation. As we seek to humanize modern science and to redefine its sociology as well as its ethical prescriptions in certain areas of inquiry, we must recognize that we are combating an aspect of its history.

The masculine metaphors commonplace in early modern science were used to great effect by its earliest progenitors. Both Bacon and Descartes spoke of science as a conqueror over a passive, "feminized" nature. The historical record seems quite clear on this point. What is not often recognized is the context within which they spoke.

Bacon sought to render science an acceptable pursuit within an aristocratic culture where war and the hunt were among the truly masculine pursuits. Descartes sought to endow science with power at a time when true masculine power was mystified by protean notions of *gloire* and the divine right of kings. Some modern critics may find little choice to be had between these old and new forms of power. But in the seventeenth century the difference between the advocacy of the power of science over nature, as distinct from that of brute force to effect the will of the possessors, must have seemed significant. Not least, our contemporary methods of social criticism rely on methods of inquiry derived from the new science of both Bacon and Descartes.

When we think as social scientists or historians we are doing so with socially focused methodologies derived from the very methodology of the new science. The invention of the social sciences, and of "scientific" history in the nineteenth century, cannot be separated historically from the method and metaphysics of early modern science, or from various scientific heresies such as materialism. However much we agree (or disagree) with Marx, Darwin, or Freud, they could not have thought as they did without the legacy of early modern science.

Much of the history here described presumes the necessity to understand culture and ideology as vital to the very fabric of historical change. The life of the mind—however much we may want to believe it to be rooted in the material order—possesses its own history, which is embodied in the written word, in language, the comprehension of which is a task of great difficulty. In the face of such difficulty it is tempting to ignore culture, to quantify the past, or to embrace the study of the material order with the expectation, commonplace since the eighteenth century, that it will always yield to our scientific inquiry. But belief in that promise of being able to dominate the material order through knowledge, as we now see, came with a host of other assumptions about nature and power, men and women, science and the state, which appear increasingly to endanger not simply our progress but also our survival. The historicity of scientific culture may, if nothing else, induce humility among those who would reduce the history of humanity to a series of mechanical laws.

In the process such reductionism offers false promises to non-Western peoples, who have every right to our historical artifacts, to our science and technology, without having to accept either our domination or our prescriptions or rules for their historical development. If the process described here serves to illustrate any principle it is that the widest possible dissemination of scientific knowledge, that is the democratization of learning, will do more to foster an indigenous creativity in matters of application or innovation than will the importation of foreign experts. The language of science must be capable of absorption, and indeed creation, by thought processes that also express other elements of the human experience. The systematic experience of nature and the codification of that experience into laws cannot be divorced from social experience. In that sense the language of science is also socially anchored, and true creativity is rooted in social experience as transformed by ingenuity.

NOTES

1. Fitzwilliam Museum, Cambridge, Strutt MS 48-1947; one of the first British industrialists, William Strutt, wrote in 1808 to Marie Edgeworth, "Ladies are excluded . . . from Mechanics and Chemistry because accurate ideas on the subject can scarcely be acquired without dirtying their persons but in other things they are competitors with the men."

BIBLIOGRAPHICAL ESSAY

Few modern studies of science and culture have approached the topic from the angle of vision used in this book, although it in turn is built on a variety of other books and essays, some general, others quite specialized. For readers wishing to pursue general topics in greater detail this bibliographical essay may be of some assistance. Yet to begin actual research on any of these subjects there is, of course, no substitute for a careful reading of the notes at the end of each chapter.

One of the major themes in this study has been the linkage between the evolution of elite culture in the early modern period and its increasing dependence on scientific thinking. As a result of that relationship we may forget the vital role played by artisanal knowledge in the earliest stages of Western scientific development. One historian pioneered that study; see the writings of E. Zilsel, for example: "The Genesis of the Concept of Scientific Progress," *Journal of the History of Ideas,* vol. 6 (1943), pp. 325–349; "The Genesis of the Concept of Physical Law," *Philosophical Review,* vol. 51 (1942), pp. 245–279; and "Problems of Empiricism: Experiment and Manual Labor," *International Encyclopedia of Unified Science* (Chicago: University of Chicago Press, 1953), vol. 2, no. 8, pp. 53–94.

There are very general questions at stake when we discuss the distinctive role of science in Western culture, but these are not always evident until we contrast our culture with another. In the Middle Ages, for example, China could be described as "far in advance" over the West in matters scientific and technical. It may not be fruitful to put the matter that simply, but the question of why that should be the case is nevertheless provocative, particularly in the light of what happened in the West from the sixteenth century onward. To address the question of Chinese science, see

Joseph Needham, *The Grand Titration* (London: Allen and Unwin, 1969), Chapters 5 and 8.

The Scientific Revolution begins with Copernicus, and his writings can be approached through Edward Rosen, *Copernicus and the Scientific Revolution* (New York: Krieger, 1984). For a recent scholarly discussion of Copernicus, see Owen Gingerich, ed., *The Nature of Scientific Discovery: A Symposium Commemorating the 500th Anniversary of the Birth of Nicolaus Copernicus* (Washington, D.C.: Smithsonian Institution, 1975). There is a general work that gives an understandable account of the relationship between Copernicus and Galileo: Clive Morphet, *Galileo and Copernican Astronomy* (London and Boston: Butterworth, 1977).

Much has been written about Galileo, and especially about his confrontation with the church. The best summary of that episode now occurs in Olaf Pedersen, "Galileo and the Council of Trent: The Galileo Affair Revisited," *Journal for the History of Astronomy*, vol. 14, no. 39 (1983). Galileo's relations with the scientific society that supported him are discussed in English in R. Morghen, "The Academy of the Lincei and Galileo Galilei," *Cahiers d'histoire mondiale*, vol. 7 (1963). An accessible short biography is Stillman Drake, *Galileo* (New York: Hill and Wang, 1980). By far the most recent and impressive work on Galileo came out at a conference whose proceedings were published in 1983: "Novità Celesti e Crisi del Sapere. Atti del Convegno Internazionale di Studi Galileiani," *Annali dell'Instituto e Museo di Storia della Scienza*. In Italian there is also Pietro Redondi, *Galileo eretico* (Turin: Einaudi, 1983).

One of the most important and fascinating topics to emerge in the study of science and culture since the 1960s is the role of magic in the new science. The *locus classicus* of those studies is Frances Yates, *Giordano Bruno and the Hermetic Tradition* (London: Routledge and Kegan Paul, 1964). Perhaps the most interesting link between magic and scientific practice occurs in early modern medicine. There the leading figure is Paracelsus. See A. G. Debus, *The English Paracelsians* (London: Oldbourne, 1965); J. Jacobi, ed., *Paracelsus: Selected Writings* (London: Routledge and Kegan Paul, 1951); and W. Pagel, *Paracelsus: An Introduction to Philosophical Medicine in the Era of Renaissance* (New York: S. Karger, 1958). For a critique of the Yates approach to the hermetic tradition and science, see R. S. Westman and J. E. McGuire, eds., *Hermeticism and the Scientific Revolution* (Los Angeles: William Andrews Clark Memorial Library, University of California, 1977), pp. 120–125.

The culture and science of the people, which increasingly came to be dismissed as magic, have been brilliantly illuminated in Keith Thomas, *Religion and the Decline of Magic* (New York, 1971); Alan Macfarlane, *Witchcraft in Tudor and Stuart England* (London, 1970); and Carlo Ginzburg, *The Cheese and the Worms* (Harmondsworth, U.K.: Penguin, 1982), about the fascinating cosmology of a miller who ran afoul of the Roman Inquisition. See also C. Ginsburg, "High and Low: The Theme of Forbidden Knowledge in the Sixteenth and Seventeenth Centuries," *Past and Present,* no. 73 (1976), pp. 28–41. And not least, to find out what ordinary folk read, see Margaret Spufford, *Small Books and Pleasant Histories: Popular Fiction and Its Readership in Seventeenth-Century England* (Athens: University of Georgia Press, 1981). The conflict between high and low culture in France is illuminated in R. Mandrou, *Magistrats et sorciers en France au 17ᵉ siècle* (Paris, 1968).

Francis Bacon is so very important in the story that links the new science to the reform of learning as well as to technology. The best places to begin with Bacon are Paolo Rossi, *Francis Bacon: From Magic to Science* (Chicago: University of Chicago Press, 1968); and B. Farrington, *The Philosophy of Francis Bacon* (Liverpool: Liverpool University Press, 1964). Bacon's influence is everywhere present in Charles Webster, *The Great Instauratian: Science, Medicine and Reform, 1626–1660* (London: Duckworth, 1975). And he is an inspiration to the founding of the Royal Society; see J. R. Jacob, "Restoration, Reformation and the Origins of the Royal Society," *History of Science,* vol. 13 (1975), pp. 155–176, which is a basic essay on the social and ideological origins of the Society.

Studies are not as plentiful as they should be on the social context of seventeenth-century French science. It is a very open field for research and study. The best place to begin, for insight on a way to read texts that anchors them within their social milieux, is with Bruce S. Eastwood, "Descartes on Refraction: Scientific Versus Rhetorical Method," *Isis,* vol. 75 (1984), pp. 481–502. See also A. J. Krailsheimer, *Studies in Self-Interest: Descartes to La Bruyère* (Oxford: Clarendon Press, 1962). There is also the now old but always valuable Martha Ornstein, *The Role of Scientific Societies in the Seventeenth Century* (Chicago: University of Chicago Press, 1928). One of the best studies on French science is Roger Hahn, *The Anatomy of a Scientific Institution: The Paris Academy of Sciences, 1666–1803* (Berkeley: University of California Press, 1971).

Both this author and J. R. Jacob have written a good deal on science and the English Revolution, and these books and essays are cited in the notes for Chapter 3. See in particular James R. Jacob, *Robert Boyle and the English Revolution* (New York: Burt Franklin, 1977), and Margaret C. Jacob, *The Newtonians and the English Revolution, 1689–1720* (Ithaca, N.Y.: Cornell University Press, 1976). A convenient summary of the general thesis at which we eventually arrived is found in James R. Jacob and Margaret C. Jacob, "The Anglican Origins of Modern Science: The Metaphysical Foundations of the Whig Constitution," *Isis,* vol. 71 (1980), pp. 251–267. Here I would like to point to new studies that extend or modify that thesis about the Anglican origins of modern science, in particular James E. Force, *William Whiston: Honest Newtonian* (Cambridge: Cambridge University Press, 1985); the student should also take a look at Richard S. Westfall, "Isaac Newton's 'Theologiae Gentiles Origines Philosophicae,' " in W. Warren Wagar, ed., *The Secular Mind . . . Essays Presented to Franklin L. Baumer* (New York: Holmes and Meier, 1982). To continue with this theme of the relationship between English science and religion, see Henry Guerlac, "Theological Voluntarism and the Biblical Analogies in Newton's Physical Thought," *Journal of the History of Ideas,* vol. 44 (1983). And there is the older essay, E. W. Strong, "Newton and God," *Journal of the History of Ideas,* no. 2 (1952), pp. 147–167. For a good warning on how difficult it can be to interpret texts written after 1660, when censorship was once again imposed, consult Steven N. Zwicker, "Language as Disguise: Politics and Poetry in the Later Seventeenth Century," *Annals of Scholarship,* vol. 1 (1980), pp. 47–67. One essay shows how even brilliant mathematicians like Hobbes could be excluded from the Royal Society for ideological reasons: Michael Hunter, "The Debate over Science," in J. R. Jones, ed., *The Restored Monarchy 1660–1688* (London 1979).

The period when science becomes a major intellectual force within Western culture can be dated as roughly 1680–1730, the so-called crisis of the European mind. For that period, the student can begin with the English translation of Paul Hazard, *The European Mind: 1680–1715* (New Haven: Yale University Press, 1953). A recent study adds to it but is rather excessively doctrinaire: Erica Harth, *Ideology and Culture in Seventeenth Century France* (Ithaca, N.Y.: Cornell University Press, 1983). There are many minor yet wonderfully fascinating historical characters that make up the story of the crisis. See, for an example, Aubrey Rosenberg, *Nicolas Gueudeville and His Work (1652–172?)* (The Hague and Boston: Nijhoff, 1982). There is also the redoubtable Henry Stubbe in England; see James R. Jacob, *Henry Stubbe: Radical Protestantism and the Early Enlightenment* (Cambridge: Cambridge University Press, 1983). One other essay takes an approach to the crisis that rightly emphasizes its relationship to the English Revolution: J. G. A. Pocock, "Post-Puritan England and the Problem of the Enlightenment" in Perez Zagorin, ed., *Culture and Politics: From Puritanism to the Enlightenment* (Los Angeles: University of California Press, 1980).

Once established as a major force in Western European culture, science in England took on an immensely practical posture that moved it from an intellectual pursuit to a source for industrialization. The classic study of that later, industrial phase is A. E. Musson and Eric Robinson, *Science and Technology in the Industrial Revolution* (Manchester: Manchester University Press, 1969). That process is at work in both England and Scotland; for the latter, see S. Shapin, "The Audience for Science in Eighteenth Century Edinburgh," *History of Science,* vol. 12 (1974), pp. 95–121, and Shapin, "Property, Patronage and the Politics of Science: The Founding of the Royal Society of Edinburgh," *British Journal for the History of Science,* vol. 7 (1974), pp. 1–41. For a good survey of eighteenth-century science in the British Isles but also in Europe, consult M. Crosland, ed., *The Emergence of Science in Western Europe* (London: Macmillan, 1975). There is also the helpful general study James E. McClellan III, *Science Reorganized: Scientific Societies in the Eighteenth Century* (New York: Columbia University Press, 1985). The larger question of science and industrial growth is tackled and somewhat downplayed in Peter Mathias, "Who Unbound Prometheus? Science and Technical Change, 1600–1800," in Peter Mathias, ed., *Science and Society* (Cambridge: Cambridge University Press, 1972). There is much more work to be done on the British literary and philosophical societies, and there are various model studies that can be imitated —for example, R. B. Schofield, *The Lunar Society of Birmingham* (Oxford: Clarendon Press, 1963); E. Robinson, "The Derby Philosophical Society," *Annals of Science*, vol. 9 (1953), pp. 359–367. Someone needs to write about the eighteenth- and early nineteenth-century engineers as the real but peculiar type of philosophes they were. There is no truly satisfactory literature to use as a guide. The history of eighteenth-century technology and culture has been largely ignored by cultural historians. But note the exception, as always: E. Robinson, "The Profession of Civil Engineer in the Eighteenth Century: A Portrait of Thomas Yeoman, F.R.S. (1704?–1781)," *Annals of Science*, vol. 18 (1962), pp. 195–216.

Scientific culture in Continental Europe during the eighteenth century needs work, and that of course requires a knowledge of various European languages. For further reading, as opposed to research, see J. L. Heilbron, *Electricity in the Seventeenth and Eighteenth Centuries: A Study of Early Modern Physics* (Berkeley: University of California Press, 1979), with a subtitle that tells more than its title. See also Peter Mathias, "Skills and the Diffusion of Innovations from Britain in the Eighteenth Century," *Transactions of the Royal Historical Society*, vol. 25 (1975). A good place to begin with French science is R. Rappaport, "Government Patronage of Science in Eighteenth Century France," *History of Science*, vol. 8 (1969), pp. 119–136. So much needs to be done on Belgian industrialization and the role of science that one hardly knows where to begin. The ability to read French and also Dutch (or Flemish as it is called in Belgium) is essential. A good general survey of the Austrian Netherlands appeared in 1983: H. Hasquin, ed., *La vie culturelle dans nos provinces au XVIIIe siècle* (Brussels: Credit Communal de Belgique). An indispensable bibliography is W. Baeten *et al.*, eds., *Belgie in de 18de eeuw: Kritische Bibliografie* (Brussels, 1983), published for the *Contact-groep 18de eeuw* and usable in French as well. Information on the eighteenth-century Dutch republic in English is even harder to find. A student can start with Margaret C. Jacob, *The Radical Enlightenment: Pantheists, Freemasons and Republicans* (London: Allen and Unwin, 1981).

The role of science in the French Revolution and the whole

question of radical science can be approached through L. P. Williams, "The Politics of Science in the French Revolution," in M. Clagett, ed., *Critical Problems in the History of Science* (Madison: University of Wisconsin Press, 1959), pp. 291–308; and R. Darnton, *Mesmerism and the End of the Enlightenment in France* (Cambridge, Mass.: Harvard University Press, 1968).

More basic research is needed on the day-to-day use of technical knowledge in the Industrial Revolution. One place to start should be on the Boulton and Watt manuscripts in Birmingham, and the book to help with that research is A. E. Musson and E. Robinson, *Science in the Industrial Revolution* (1969), cited earlier. This material can be approached in the shorter essay by the same authors, "Science and Industry in the Late Eighteenth Century," *Economic History Review*, 2nd ser., vol. 13 (1960–1961), pp. 222–244. A case study of one of the new sciences and its relation to industrialization is R. Porter, "The Industrial Revolution and the Rise of the Science of Geology," in M. Teich and R. M. Young, eds., *Changing Perspectives in the History of Science* (London: Heinemann, 1973), pp. 320–343; and see also A. Thackray, "Science and Technology in the Industrial Revolution," *History of Science*, vol. 9 (1970), pp. 76–89. And finally a good general text on science per se in the eighteenth century has now been published: Thomas Hankins, *Science and the Enlightenment* (Cambridge: Cambridge University Press, 1985).

There is now also a textbook for history and social science students that helps to explain the approach taken to science by the new methodology. See John Law and Peter Lodge, *Science for Social Scientists* (London: Macmillan, 1984; U.S. distribution, Humanities Press, Atlantic Highlands, N.J.) There is also a useful course description of the kind of social history of science course that is now being taught at the University of Edinburgh, but which could be adapted: Steven Shapin, "A Course in the Social History of Science," *Social Studies of Science*, vol. 10, no 2 (1980), pp. 231–258. If students wish to know about individual scientists discussed in this text, they should consult Charles C. Gillispie, ed., *Dictionary of Scientific Biography*, 16 vols. (New York: Scribner, 1970). For complex ideas in philosophy, there is the helpful guide by Philip P. Wiener, ed., *Dictionary of the History of Ideas* (New York: Scribner, 1973).

GLOSSARY OF TERMS

Anabaptists Originating in Germany and the Low Countries, these Protestant sectaries of the 1520s and beyond practised adult baptism, or rebaptism, and believed that only the specially illuminated could choose salvation.

Anglican church The official and legally established Church of England; its place was guaranteed by an act of Parliament during the Henrican Reformation of the 1530s. It was disestablished by Parliament during the English Revolution and reinstated at the restoration of the monarchy in 1660.

Arminians Liberal Calvinists of Dutch origin who rejected a rigid version of predestination and who, thereby, emphasized free will. They frequently favored the power of lay magistrates over the clergy and hence opposed a theocratic (i.e., independent) political role for the clergy.

atomism An ancient doctrine that asserted the existence of minuscule, hard, indivisible particles as the essence of the material order. The ancient Greek poet Lucretius was a common source of atomism, which he in turn had learned from Epicurus. The doctrine was revived in the early seventeenth century, often as an alternative to Aristotelian notions of matter and form as the key to the physical reality we observe around us. The French natural philosopher Gassendi (see p. 47–51) attempted to Christianize this pagan philosophy by arguing that God directs the atoms.

corpuscular philosophy A term used to describe a natural philosophy based on an essentially atomistic understanding of matter. The term is most commonly associated with the English scientist Robert Boyle, whose philosophy was also deeply Christian.

Dissenters Non-Anglican English Protestants, who could be Presbyterians, Congregationalists, Quakers, Baptists, and so on. They possessed religious freedom in the eighteenth century but were still barred from the universities and from holding political office.

Erastian A term used to describe those who would subordinate the church to the state.

Hanoverians George I, elector of Hanover, became king of England in 1714 as a result of a parliamentary bill that offered him the throne because he was a Protestant, although a German, cousin of the Stuarts. He and his successors in the eighteenth century, George II and George III, are frequently referred to as the Hanoverian kings.

Huguenots The name given to French Protestants. After 1685 and the revocation of the Edict of Nantes, and hence the end of religious toleration in France, many Huguenots became refugees in The Netherlands, Switzerland, and England.

Jacobinism A term referring to the republican and democratic ideology of the Jacobins of the French Revolution; Robespierre was their most famous representative.

latitudinarianism A term first used in the 1660s ("latitude-man") to denote someone who favored the restoration of the Anglican church, but with some accommodation to Dissenters. Both clergymen and laymen were attracted to this moderate and liberal version of Anglicanism, which was also supportive of the new science.

philosophe A term applied in the eighteenth century and beyond to the advocates and promoters of the Enlightenment. The philosophes wrote in a popular style and eschewed the complex forms of argumentation commonly used by philosophers. On the Continent they were frequently anticlerical and attracted to deism, occasionally to materialism.

Puritans The name given in the early seventeenth century to English Protestants who wanted to "purify" the English church by instituting an ecclesiastical government more responsive to the laity than they saw the bishops as being, and also to purge older Catholic rituals from the ceremonies of the church. They frequently looked to Calvin's Geneva as the model for church government and ceremonial purity as well as for the source of doctrinal orthodoxy.

royalist Anyone who supports the authority of kings. In England during the revolution this meant someone who supported the Stuarts (namely, Charles I, executed in 1649) and advocated their restoration (as occurred in 1660 with Charles II).

schoolmen A term used to denote clergymen who taught in the advanced schools and universities of early modern Europe and who generally subscribed to the version of Aristotle found in the medieval writings of Thomas Aquinas and his commentators.

transubstantiation The belief in the Real Presence of Christ in the sacrament of the Eucharist, that is, in an actual transformation of the bread and wine consecrated by the priest during a mass. The doctrine of consubstantiation, first put forward by Protestant reformers, would permit both the bread and the body of Christ to be present.

Unitarians Eighteenth-century proponents of rational religion who rejected the doctrine of the Trinity and had little affiliation with traditional religiosity.

virtuosi A term used in England to describe natural philosophers or, after 1662, simply members of the Royal Society.

Whigs A political party formed in England in the early 1680s as a result of the unsuccessful attempt to exclude James II from the throne because he was a Catholic. After the Revolution of 1688–1689 the party came to power within court circles and championed the interests of the commercial and landed oligarchy. It was bitterly opposed by the Tory party, which would frequently win parliamentary elections but whose members seldom obtained court or ministerial power.

NAME INDEX

Academy of Science, 63–64, 188, 202–4
Albury, W.R., 94
Alcock, Nathan, 150
Algarotti, F., 207
Allamand, J.N.S., 119
Amsterdam, 7, 35, 108, 127–8, 187, 189, 191–2, 195–6
Anabaptists, 27, 189, 262
Aristotle, 3, 13, 17–19, 21, 28, 32, 44, 62, 87–88, 110
Aristotelianism, 12
Arkwright, Sir Richard, 131, 200, 239
Athenaeum, 196–7
Avon, River, 226, 230

Bacon, Francis, 20, 31–33, 63, 73, 76, 113, 131, 156, 169, 202, 204, 252, 257
Banks, Sir Joseph, 155, 244
Barker, John, 243
Barneveld, Willem van, 197
Barrow, Isaac, 96
Baxter, Richard, 84
Bayle, Pierre, 113, 121
Beeckman, Isaac, 35, 51–52, 182, 189
Bekker, Balthasar, 115, 119
Belgium, 130, 179, 184, 190, 198
Bentley, Richard, 89, 96, 123, 141–2
Bentley, Thomas, 136
Bible, the, 18, 21, 23, 28, 82, 113, 116, 119, 122
Birmingham, 164, 166, 202, 237
Boerhaave, Herman, 36, 125, 185

Bologna, 16
Booth, John, 144
Bosma, Benjamin, 195–6, 207
Boulton, Matthew, 166, 239
Boyle, Robert, 73, 75, 77–78, 81, 83–85, 107–8, 111, 114, 124, 138, 193, 206
Bright, Richard, 224, 231
Bristol, 223
Britain, 184, 190, 192
British Association for the Advancement of Science, 36
Bruno, Giordano, 14, 26, 44, 80, 117
Brussels, 118, 128
Burnett, Gilbert, 108

Calabria, 28
Cambon-van Werken, Marie Gertruide de, 151
Cambridge, 49, 77, 86, 88, 92, 150
Campanella, Thomas (1568–1639), 28–29
Chambers, Ephraim, 154
Champion, William, 230
Charleroi, 199
Charles I, 77
Charles II, 92
Chester, 158
China, 255
Ciampoli, Giovanni, 22
Clare, Martin, 146, 154
Clarke, Samuel, 96–97, 122–4, 141–2, 185
Colbert, 63–65, 67, 202
College of William and Mary, 166

SUBJECT INDEX

absolutism, 59, 61–62, 65, 77,
91–92, 106–07, 112; attacks
on, 109–10, 119; and
Cartesianism, 116;
enlightened, 202; in Spain,
107–08; threat of, 107–09
academies, of Dissenters, 130,
146–47, 150, 201–02, 231
agriculture, 157, 207–08
alchemy, 35, 58, 90, 115
American colonies, 128, 165
Anglican church, 30, 32, 76, 89,
91, 93, 166; hegemony of,
138; and science, 69, 73–74,
84 *passim*, 96, 150
animism, 82
antichrist, 113
antinomianism, 82
Aristotelianism, see scholasticism
armies, 8, 58, 79, 107, 205–06
Arminians, Dutch, 122, 262; see
also Le Clerc, Jean
artisans, 32, 37, 76–77; scientific
education for, 146–47,
168–69
astrology, 22, 35, 47, 113, 223
astronomy, 17, 21, 24, 63, 198
atheism, 53, 67, 69, 80, 89, 120–21
atomism, 48–49, 53, 81, 87–88,
108, 206, 262
Austrian Netherlands, see
Belgium

books, 28, 76, 130, 153, 191; see
also printing
bourgeoisie (of Paris), 47
Boyle lectures, 96–97, 111,
141–42, 175
Britain, see England

Calvinism, Dutch, 52–53, 68,
114–15, 122, 186, 189
capitalism, 66, 83, 121, 162, 167,
193–94, 222, 231; and
industrialization, 183–84
canal building, 162, 208, 221–22,
227, 230, 239–40
Cartesianism, 25, 49, 54 *passim*,
85, 116, 182–83; repudiation
of, 86–87; at Sorbonne,
201–02
censorship, 64, 88, 106–07, 183;
see also books
chemistry, 57, 66, 81, 193, 201;
and industry, 199; see also
alchemy
Church of England, see Anglican
church
Church, the Roman Catholic, 22,
25, 33, 85, 87, 116; and
Counter Reformation, 26
clergy, 7, 11–12, 27, 32, 48, 51,
83, 107, 113, 179, 189; in
Belgium, 199; in France, 62,
67; and freemasonry, 127;
vs. Galileo, 17, 24; and
Hobbes, 81; in The
Netherlands, 52–53, 67–68,
116, 189; vs. new science,
120, 144; and Whigs, 123
coal mining, 129–30, 140, 154,
190, 227
corresponding societies, 165
cotton mills, 159, 200
courts (of kings), 8, 11, 58, 96
crisis (of the late seventeenth
century) 68, 105 *passim*,
128
culture, high vs. low, 19

ABOUT THE AUTHOR

MARGARET C. JACOB is currently Dean of Eugene Lang College and Professor of History in the University at the New School for Social Research. She received her Ph.D. from Cornell University, and has taught in British, Dutch, and American universities. Dr. Jacob has been a research fellow at The Institute for Advanced Study in Princeton, Harvard University, and a Fulbright scholar at the University of Leiden. Among her previous publications are three books, two edited collections, numerous articles covering diverse topics in European cultural history, with emphasis on the social context of science, the relationship between science and religion, the formation of political ideologies, freemasonry, and women's history. She has chaired the program committee of the American Historical Association, co-founded the Institute for Research in History, and has been the recipient of fellowships from the National Endowment for the Humanities, the American Council of Learned Societies, and the National Science Foundation. When not being dean, she continues to teach and write.

A NOTE ON THE TYPE

This book was set in a *digitized version* of Baskerville, originally a recutting of a typeface designed by John Baskerville (1706–1775). Baskerville, a writing master in Birmingham, England, began experimenting about 1750 with type design and punch-cutting. His first book, set throughout in his new types, was a Virgil in royal quarto published in 1757, and this was followed by other famous editions from his press. Baskerville's types were a forerunner of what we know today as the "modern" group of typefaces.